MyMathLab®

Notebook

Gex, Incorporated

Introductory Algebra

John Squires
Chattanooga State Community College

Karen Wyrick
Cleveland State

PEARSON

Boston Columbus Indianapolis New York San Francisco Upper Saddle River
Amsterdam Cape Town Dubai London Madrid Milan Munich Paris Montreal Toronto
Delhi Mexico City Sao Paulo Sydney Hong Kong Seoul Singapore Taipei Tokyo

Copyright © 2012 Pearson Education, Inc.
Publishing as Pearson, 75 Arlington Street, Boston, MA 02116.

Bound: ISBN-13: 978-0-321-74926-0
            ISBN-10: 0-321-74926-X
Unbound: ISBN-13: 978-0-321-78614-2
            ISBN-10: 0-321-78614-9

1 2 3 4 5 6 BRR 15 14 13 12 11

www.pearsonhighered.com

MyMathLab, Introductory Algebra: A Modular Approach

John Squires and Karen Wyrick

Table of Contents

Name: _____   Date: _____

Instructor: _____   Section: _____

Real Numbers and Variables
Topic 1.1 Introduction to Real Numbers

Vocabulary

whole numbers   •   integers   •   rational number
irrational number   •   real numbers   •   radical sign

1. _____ are all the rational numbers and all the irrational numbers.

2. The decimal form of a(n) _____ is non-terminating and non-repeating.

Step-by-Step Video Notes

Watch the Step-by-Step Video lesson and complete the examples below.

Example	Notes
1. Classify each number as an integer, a rational number, an irrational number, and/or a real number. Circle all that apply. 8 integer rational irrational real $-\dfrac{1}{5}$ integer rational irrational real $\sqrt{3}$ integer rational irrational real	
2–4. Plot the following real numbers on a number line. 1 -2.75 $\dfrac{1}{2}$	

1

Example	Notes
5–7. Use a real number to represent each real-life situation. A temperature of 131.2 °F below zero is recorded in Antarctica. The height of a mountain is 22,645 feet above sea level. A golfer scored 5 under par in a recent tournament.	

Helpful Hints

Rational numbers are any numbers that can be written as fractions in the form $\frac{a}{b}$, where a and b are integers. The decimal equivalents of rational numbers repeat or terminate.

Irrational numbers, such as π or $\sqrt{2}$, are non-terminating, non-repeating decimals.

Positive numbers are to the right of 0 on a number line. Negative numbers are to the left of 0 on a number line.

Concept Check

1. You would plot $-\frac{7}{2}$ between what two integers on a number line?

Practice

Classify each number as rational or irrational.

2. 0.3333...

3. $\sqrt{16}$

Use a real number to represent the situation.

4. New Orleans is 64 feet below sea level.

5. Kelsey deposits $125 into her savings account.

2

Real Numbers and Variables
Topic 1.2 Graphing Real Numbers Using a Number Line

Vocabulary
number line   •   inequality   •   inequality symbols

1. A(n) _____ is a statement that shows the relationship between any two real numbers that are not equal.

2. The _____ ">" and "<" are used to represent the phrases " is greater than" and "is less than."

Step-by-Step Video Notes
Watch the Step-by-Step Video lesson and complete the examples below.

Example	Notes
1 & 2. Plot the following real numbers on a number line. 2 ![number line from -3 to 3] -3 -2 -1 0 1 2 3 -1.5 ![number line from -3 to 3] -3 -2 -1 0 1 2 3	
3 & 4. Write the following statements using inequality symbols. 2 is less than 7 8 is greater than 5	

Example	Notes

Example

5–7.   Plot the given numbers on a number line, and then replace the question mark with the appropriate symbol, > or <.

−2.5 ? 0   −2.5 ☐ 0

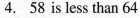

$4 \ ? \ \dfrac{3}{4}$ $4 \ \boxed{} \ \dfrac{3}{4}$

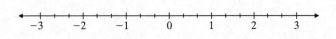

−1 ? −5   −1 ☐ −5

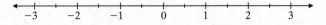

Helpful Hints

When comparing numbers, one number is greater than another number if it is to the right of that number on the number line. One number is less than another number if it is to the left of that number on the number line.

Remember that the inequality symbols ">" and "<" point toward the lesser value.

Concept Check

1. How can you tell which is the greater of two numbers plotted on a number line?

Practice

Plot the real numbers on a number line.

2. $2\dfrac{2}{3}$

Write statements using inequality symbols.

4. 58 is less than 64

3. −0.5

5. −7 is greater than −10

4

Real Numbers and Variables
Topic 1.3 Translating Phrases into Algebraic Inequalities

Vocabulary
variable   •   inequality phrase   •   inequality symbols

1. A(n) _____ is a letter or symbol that is used to represent an unknown quantity.

Step-by-Step Video Notes
Watch the Step-by-Step Video lesson and complete the examples below.

Example	Notes
1. Translate the phrase into an algebraic inequality. A police officer claimed that a car was traveling at a speed more than 85 miles per hour. (Use the variable *s* for speed.) Determine the phrase to be translated. Replace the unknown quantity with the given variable. Replace the inequality phrase with the correct inequality symbol. Answer:	
3. Translate the phrase into an algebraic inequality. The owner of a trucking company said that the payload of a truck must be no more than 4500 pounds. (Use the variable *p* for payload.) Determine the phrase to be translated. Replace the unknown quantity with the given variable. Answer:	

Example	Notes
4. Translate the phrase into an algebraic inequality.	

Carlos must be at least 16 years old in order to get his driving license. (Use the variable *a* for age.)

Determine the phrase to be translated.

Replace the unknown quantity with the given variable.

Answer:

Helpful Hints

When translating phrases into algebraic inequalities look for key words or phrases to determine which inequality symbol is most appropriate to use.

The inequality symbol "$\leq$" can also mean "at most" or "no more than." The inequality symbol "$\geq$" can also mean "at least" or "no less than."

Concept Check

1. Is $-5 \leq -5$? Is $\frac{7}{4} \geq 1\frac{3}{4}$?

Practice

Translate each phrase into an algebraic inequality.

2. According to the building inspector, the elevator can hold at most 12 people. (Use the variable *n* for number of people.)

3. A truck must be no taller than 9.5 feet to go through a certain tunnel. (Use the variable *h* for the height of the truck.)

4. You must be at least 54 inches tall to go on the roller coaster. (Use the variable *i* for the number of inches.)

Real Numbers and Variables
Topic 1.4 Finding the Absolute Value of a Real Number

Vocabulary
distance   •   number line   •   opposites   •   absolute value

1. The _____ of a number is the distance between that number and zero on a number line.

Step-by-Step Video Notes
Watch the Step-by-Step Video lesson and complete the examples below.

Example	Notes
1 & 2. Use a number line to find the absolute value of the following numbers. 3 $\begin{array}{ccccccc} -3 & -2 & -1 & 0 & 1 & 2 & 3 \end{array}$ −1.5 $\begin{array}{ccccccc} -3 & -2 & -1 & 0 & 1 & 2 & 3 \end{array}$	
3. Write the expression for the absolute value of $-\dfrac{3}{8}$. _____	
4–6. Find the absolute value of the following numbers. $\left\lvert -3.68 \right\rvert = \boxed{}$ $\left\lvert 3.68 \right\rvert = \boxed{}$ $\left\lvert 0 \right\rvert = \boxed{}$	

Example	Notes
7–9. Find the absolute value of the following numbers.	

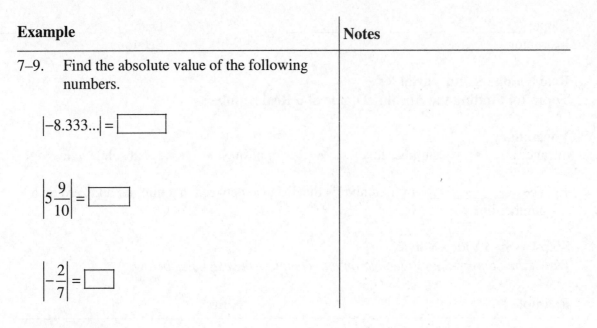

$$|-8.333...| = \boxed{}$$

$$\left|5\frac{9}{10}\right| = \boxed{}$$

$$\left|-\frac{2}{7}\right| = \boxed{}$$

Helpful Hints

The absolute value of a number can be thought of as the numerical part of the number, without regard to its sign.

Absolute value is the distance from zero on a number line. Since it represents a distance, absolute value can never be negative.

Concept Check

1. How many numbers have an absolute value of 6.5? of $-\frac{3}{4}$?

Practice

Write the expression for the absolute value of the numbers given.

2. −.8888...

Find the absolute value of the following numbers.

4. $|-24|$

3. $\frac{1}{4}$

5. $\left|-6\frac{7}{8}\right|$

Adding and Subtracting with Real Numbers
Topic 2.1 Adding Real Numbers with the Same Sign

Vocabulary
absolute values • positive numbers • negative numbers • same sign

1. When adding add two numbers of the _____, add the absolute values, or numerical parts, of the numbers and give the answer the same sign as the numbers being added.

Step-by-Step Video Notes
Watch the Step-by-Step Video lesson and complete the examples below.

Example	Notes
1 & 2. Use chips to add the following. $4+2$ $\oplus \oplus \oplus \oplus + \oplus \oplus = \oplus \oplus \oplus \oplus \oplus \oplus$ Answer: $-2+(-3)$ $\odot \odot + \odot \odot \odot = \odot \odot \odot \odot \odot$ Answer:	
4. Add $-3+(-5)$. Add the numerical parts. Give the answer the sign of the numbers being added. Answer:	

Example	**Notes**
5. Add $-\dfrac{2}{3}+\left(-\dfrac{1}{7}\right)$. Add the numerical parts. Find a common denominator. Answer:	

6. Add $-8.1+(-2.75)+(-5.03)$.

Add from left to right.

Answer:

Helpful Hints
To add numbers with the same sign, add the numerical parts of the numbers and give the answer the same sign as the numbers being added.

A number written without a sign is assumed to be positive.

Concept Check
1. What will be the sign of the sum of $-\dfrac{3}{5}+\left(-\dfrac{4}{9}\right)$?

Practice
Add.

2. $6.27+12.8$

4. $-8+(-8)$

3. $-\dfrac{4}{5}+\left(-\dfrac{7}{10}\right)$

5. $-8.43+(-0.57)$

Adding and Subtracting with Real Numbers
Topic 2.2 Adding Real Numbers with Different Signs

Vocabulary
different signs   •   numerical part   •   common denominator

1. When adding add two numbers with _____, subtract the absolute values, or numerical parts, of the numbers and give the answer the same sign as the larger numerical part.

Step-by-Step Video Notes
Watch the Step-by-Step Video lesson and complete the examples below.

Example	Notes
1 & 2. Use chips to add the following. $6+(-4)$ $\oplus\ \oplus\ \oplus\ \oplus\ \oplus\ \oplus$ $+\odot\ \odot\ \odot\ \odot$ Answer: $-7+3$ $\odot\ \odot\ \odot\ \odot\ \odot\ \odot\ \odot$ $+\oplus\ \oplus\ \oplus$ Answer:	
3. Add $-6+9$. Subtract the numerical parts. Give the answer the same sign as the larger "numerical part." Answer:	

Example	Notes
4. Add $\dfrac{3}{7}+\left(-\dfrac{5}{7}\right)$. Answer:	

5. Add $3.7+(-10.5)$.

Answer:

Helpful Hints

Remember that the absolute value of a number is the distance between that number and zero on a number line.

Try to determine the sign of the answer before calculating. You're less likely to forget to give the answer the correct sign.

Concept Check

1. What will be the sign of the sum of $5+\left(-3\dfrac{2}{9}\right)$?

Practice
Add.

2. $6+(-13)$

4. $-7.7+8.7$

3. $\dfrac{4}{7}+\left(-\dfrac{2}{7}\right)$

5. $-\dfrac{7}{8}+\dfrac{2}{3}$

Adding and Subtracting with Real Numbers
Topic 2.3 Finding the Opposite of a Real Number

Vocabulary
opposite numbers   •   additive inverses   •   absolute value

1. Two numbers that differ only in sign are _____. They are the same distance from zero on a number line but in opposite directions.

2. Two numbers are _____, or opposites, if they add to equal zero.

Step-by-Step Video Notes
Watch the Step-by-Step Video lesson and complete the examples below.

Example	**Notes**
1. Find the opposite of the situation. A temperature increase of 8.5 °F. The opposite of increase is _____. The opposite of the situation is a _____ of 8.5 °F.	
3–5. Find the opposite of the following numbers. $7 \qquad -3.5 \qquad -\dfrac{5}{7}$ To find the opposite of a number, change the sign of the number. The opposite of 7 is $\boxed{}$. The opposite of −3.5 is $\boxed{}$. The opposite of $-\dfrac{5}{7}$ is $\boxed{}$.	

Example	Notes
6 & 7. Find the opposite of the following absolute values. $$\|5\| \qquad \|-6.78\|$$ $$-\|5\| = \boxed{}$$ $$-\|-6.78\| = \boxed{}$$	

8. Find the additive inverse, or opposite, of -4. Then add the additive inverse to the number.

The additive inverse of -4 is $\boxed{}$.

$$-4 + \left(\boxed{} \right) = \boxed{}$$

Helpful Hints

For any real number a, $-(-a) = a$, but $-\|-a\| = -a$.

For any real number, its opposite and its additive inverse are the same number.

Concept Check

1. What will be the sign of the additive inverse of $\left(-6\frac{7}{8} \right)$?

Practice

Find the additive inverse, or opposite, of each number.

2. -2.3

3. $-\left(-\dfrac{4}{5} \right)$

Find the opposite of the following absolute values.

4. $\|4.4\|$

5. $\left\| -\dfrac{2}{3} \right\|$

Adding and Subtracting with Real Numbers
Topic 2.4 Subtracting Real Numbers

Vocabulary
difference   •   add the opposite   •   sum

1. To subtract real numbers, _____ of the second number to the first. This sometimes is referred to this as leave, change, change.

Step-by-Step Video Notes
Watch the Step-by-Step Video lesson and complete the examples below.

Example	Notes
1. Subtract $10 - 40$. Leave the first number alone. Change the minus sign to a plus sign. Change the sign of the number being subtracted. Add the two numbers using the rules of addition. Answer:	
3. Subtract $-5 - (-9)$. Answer:	

Example	Notes
4. Subtract $-6.18 - 2.34$. Answer:	
5. The temperature in a city is $-8\,^\circ$F and the temperature of a neighboring town is $-15\,^\circ$F. Find the difference in temperature between the two cities. Write the difference as a subtraction problem. Answer:	

Helpful Hints

Subtracting a number is the same as adding its opposite.

Any subtraction problem with real numbers can be changed to an addition problem by using the methods "Add the Opposite" or "Leave, Change, Change."

Concept Check
1. What will always be the sign of the difference when you subtract a positive number from a negative number?
2. What will always be the sign of the difference when you subtract a negative number from a positive number?

Practice

Subtract. Use "Add the Opposite."

3. $-2 - (-9)$

4. $\dfrac{4}{5} - \left(-\dfrac{3}{5}\right)$

Subtract. Use "Leave, Change, Change."

5. $-2.7 - 5.4$

6. $5\dfrac{1}{3} - 6\dfrac{1}{2}$

Name: _____   Date: _____

Instructor: _____   Section: _____

Adding and Subtracting with Real Numbers
Topic 2.5 Addition Properties of Real Numbers

Vocabulary
Commutative Property of Addition   •   Addition Property of Zero   •   Associative
Property of Addition   •   Additive Inverse Property   •   Distributive Property

1. The _____ states that changing the order of the numbers being added
    does not change the sum.

2. The _____ states that changing the grouping when adding numbers does
    not change the sum.

3. In symbols, the _____ states that for any real number a, $a + (-a) = 0$
    and $-a + a = 0$.

Step-by-Step Video Notes
Watch the Step-by-Step Video lesson and complete the examples below.

Example	Notes
Write equivalent expressions using the commutative property of addition. $$3 + 6 = \square + \square$$ $$-7 + (-8) = \square + (\square)$$	
Write equivalent expressions using the addition property of zero. $$0 + 5 = \square$$ $$-8 + 0 = \square$$	
Write an equivalent expression using associative property of addition. $$-6 + (-4 + 2) = \boxed{}$$	

Example	Notes
1–5. Determine which property of addition is exhibited by each equation. The properties are Commutative, Associative, Addition Property of Zero, and Additive Inverse Property.	

$0 + (-7) = -7$ _____

$(-2 + 3) + 5 = 5 + (-2 + 3)$ _____

$10 + (-10) = 0$ _____

$(-2 + 6) + 0.3 = -2 + (6 + 0.3)$ _____

$-\dfrac{2}{3} + \dfrac{1}{3} = \dfrac{1}{3} + \left(-\dfrac{2}{3}\right)$ _____

Helpful Hints

Use the properties of addition in cases where a property may make it easier for you to simplify an expression.

Concept Check

1. Which property of addition will make it easier for you to simplify the expression $6.2 + (3.8 + (-7.2))$? Explain how you would use this property in this case.

Practice

Match each description with the addition property that it describes.

2. This addition property deals with changing the order of the numbers being added.

Write an equivalent expression using the Associative Property of Addition.

4. $5.2 + (-5.2 + (-7.2))$

3. This addition property deals with adding a number to its opposite.

5. $(-8.6 + 5) + 9.03$

Multiplying and Dividing with Real Numbers
Topic 3.1 Multiplying Real Numbers

Vocabulary

absolute values   •   product   •   negative factors

1. When multiplying two numbers with the same sign, the _____ will always be positive.
2. When multiplying an odd number of _____, the product is negative.

Step-by-Step Video Notes
Watch the Step-by-Step Video lesson and complete the examples below.

Example	Notes
1. Multiply $(7)(-4)$. Multiply the numerical parts of the numbers. $(7)(4) = \boxed{}$ Determine the sign of the answer. $(+)(-) = \boxed{}$ Answer:	
3. Multiply $(-12)(-9)$. Answer:	
4. Multiply $(-3.5)(-2)$. Answer:	

Example	Notes
5–8. Multiply.	

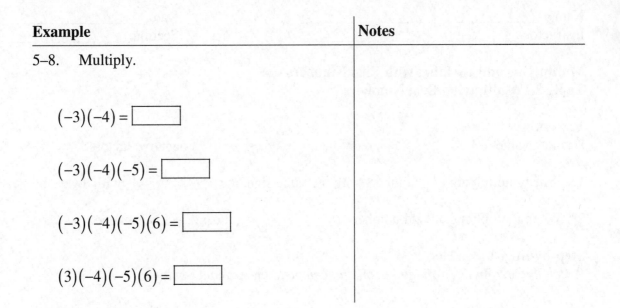

$$(-3)(-4) = \boxed{}$$

$$(-3)(-4)(-5) = \boxed{}$$

$$(-3)(-4)(-5)(6) = \boxed{}$$

$$(3)(-4)(-5)(6) = \boxed{}$$

Helpful Hints

When multiplying two numbers with the same sign, the product is positive. When multiplying two numbers with different signs, the product is negative.

When multiplying an even number of negative factors, the product is positive. When multiplying an odd number of negative factors, the product is negative.

Concept Check

1. Will the product of $-(-27)(-132)$ be positive or negative? Explain.

Practice
Multiply.

2. $(-0.8)(5)$ 4. $(-2)(-5)(-3)$

3. $\left(-\dfrac{1}{4}\right)\left(-\dfrac{2}{3}\right)$ 5. $(-2)(-5)(-3)(-8)$

Name: _____   Date: _____

Instructor: _____   Section: _____

Multiplying and Dividing with Real Numbers
Topic 3.2 Finding the Reciprocal of a Real Number

Vocabulary
integer   •   reciprocals   •   multiplicative inverse   •   invert

1. Two numbers are _____ of each other if their product is 1.

2. The reciprocal is also called the _____ of a number.

Step-by-Step Video Notes
Watch the Step-by-Step Video lesson and complete the examples below.

Example	Notes
1. Find the reciprocal of $-\dfrac{5}{7}$. The number is already written as a fraction. Invert the fraction. The sign of the reciprocal is the same as the sign of the original number. Answer:	
2. Find the reciprocal of -3. Write the number as a fraction with 1 as the denominator. Invert the fraction. Answer:	

Example	Notes
3. Find the reciprocal of 0.5 . Write the number as a fraction with 1 as the denominator. Answer:	
4. Find the reciprocal of $-4\dfrac{3}{5}$. Write the number as an improper fraction. Answer:	

Helpful Hints

A negative fraction can be written in 3 different but equal ways: $-\dfrac{a}{b} = \dfrac{-a}{b} = \dfrac{a}{-b}$.

To find the reciprocal of an integer or a decimal, start by putting the number over 1.

To find the reciprocal of a mixed number, first write the number as an improper fraction.

Concept Check
1. What is the sign of the reciprocal of a negative number? A positive number?

Practice
Find the reciprocal.

2. $-\dfrac{3}{8}$ 4. $-5\dfrac{1}{3}$

3. 4   5. 0.4

Multiplying and Dividing with Real Numbers
Topic 3.3 Dividing Real Numbers

Vocabulary
division   •   quotient   •   dividend   •   divisor   •   mixed number

1. The operation of splitting a quantity or number into equal parts is_____.

2. The number you are dividing by in a division problem is called the _____.

3. When dividing numbers with different signs, the _____ will be negative.

Step-by-Step Video Notes
Watch the Step-by-Step Video lesson and complete the examples below.

Example	Notes
1–3. Divide. $(-21) \div (-7)$ $3.2 \div 0.8$ $\dfrac{-45}{25}$	
4. Four friends decide to start a business together. They share a start-up loan of $\$120,000$. If they split the amount of the loan equally between them, how much does each friend owe? The loan can be represented as a negative number. $(-120,000) \div 4 = \boxed{}$ Answer:	

Example	Notes
5 & 6. Divide. $32 \div \left(-\dfrac{8}{3} \right)$ $\left(-\dfrac{11}{6} \right) \div 2\dfrac{4}{9}$	

Helpful Hints

The fraction bar in $\dfrac{a}{b}$ is another way to express division. The expression $\dfrac{a}{b}$ is the same as $a \div b$. Follow the rules for division to simplify.

When dividing by a fraction, find the reciprocal of the divisor, and multiply. If the divisor is a mixed number, rewrite it as an improper fraction and divide as stated above.

Concept Check

1. Rewrite the division problem $\left(\dfrac{-21}{25} \right) \div (-7)$ to show the equivalent multiplication by the reciprocal of the divisor.

Practice
Divide.

2. $-56 \div (-7)$

3. $\left(\dfrac{-12}{35} \right) \div \left(-\dfrac{6}{7} \right)$

4. $12.8 \div (-0.8)$

5. $14 \div \left(-1\dfrac{3}{4} \right)$

Multiplying and Dividing with Real Numbers
Topic 3.4 Exponents and the Order of Operations

Vocabulary
base   •   exponent   •   order of operations   •   even power

1. An exponent is used as a shortcut for repeated multiplication. The _____
   is the number being multiplied.

Step-by-Step Video Notes
Watch the Step-by-Step Video lesson and complete the examples below.

Example	Notes
1. Write in exponential form. $\left(-\dfrac{5}{7}\right)\left(-\dfrac{5}{7}\right)\left(-\dfrac{5}{7}\right)\left(-\dfrac{5}{7}\right)$ Answer:	
4. Evaluate. $\left(-\dfrac{2}{5}\right)^3$ Answer:	
7 & 8. Evaluate. $(-5)^2$ $-\left(-\dfrac{1}{5}\right)^3$	

Example	Notes
9 & 10. Evaluate.	

$$\frac{4^3 + 2(-5)}{2^3}$$

$$(-4)^2 - 2(5)^2$$

Helpful Hints

Any non-zero number raised to the zero power is equal to 1. Any number raised to the 1 power is equal to itself.

A negative base raised to an even power is positive. A negative base raised to an odd power is negative. Notice that $(-2)^4 = 16$, while $-2^4 = -16$.

The fraction bar acts like parentheses when evaluating an expression. Find the value of the expression in the numerator and the value of the expression in the denominator, then divide.

Concept Check

1. What is the sign of the product $(-9)(-7)^3$?

Practice

Evaluate.

2. $\left(-\frac{1}{4}\right)^1$

3. $(456789321)^0$

4. $\dfrac{-6 + (-5)^2}{-(-2)^3 + 3}$

5. $(-0.4)^2 + (0.3)^2 - (0.25)^1$

Multiplying and Dividing with Real Numbers
Topic 3.5 The Distributive Property

Vocabulary
mental multiplication   •   distributive property   •   order of operations

1. The _____ states that for all real numbers a, b, *and* c,
   $a(b+c) = ab + ac$.

Step-by-Step Video Notes
Watch the Step-by-Step Video lesson and complete the examples below.

Example	Notes
1. Multiply $9 \cdot 103$. Rewrite as an addition problem. $9 \cdot 103 = \Box (\boxed{} + \Box)$ Use the Distributive Property to multiply, and then simplify. $9(100) + (\Box)(\Box)$ Answer:	
3. Multiply $-3(-8x+3)$. Rewrite using the Distributive Property. $-3(-8x+3) = (-3)(-8x) + (\Box)(\Box)$ Simplify the result. $(-3)(-8x) + (-3)(3) = \boxed{}$ Answer:	

Example	Notes

4–7.   Multiply.

$2(-4y+7) = \boxed{}$

$-9(-4x-6) = \boxed{}$

$-(x-2) = \boxed{}$

$5(3x+2y-6z) = \boxed{}$

Helpful Hints

Use the Distributive Property to make mental multiplication easier whenever you can change one of the factors into a sum where one of the addends is an easy number to multiply.

The Distributive Property is important and useful in simplifying algebraic expressions when you cannot perform the operations inside parentheses. To remove the parentheses, be sure to multiply by each term inside the parentheses.

Concept Check

1.   What must you multiply each term in the parentheses by to simplify $-(x+y-4)$?

Practice

Rewrite using the Distributive Property.

2. $9\left(-\dfrac{1}{3}+\dfrac{2}{9}\right)$

3. $104(-8)$

Simplify using the Distributive Property.

4. $-6(-5x+4)$

5. $9(-3x+4y-7z)$

Multiplying and Dividing with Real Numbers
Topic 3.6 Multiplication Properties of Real Numbers

Vocabulary
Commutative Property   •   Zero Property   •   Associative Property
Identity Property

1. The _____ of Multiplication states that changing the order when multiplying numbers does not change the product.

2. The _____ of Multiplication states that changing the grouping when multiplying numbers does not change the product.

Step-by-Step Video Notes
Watch the Step-by-Step Video lesson and complete the examples below.

Example	Notes
1–4. Determine which property of multiplication is shown by each equation: the Commutative Property of Multiplication, the Associative Property of Multiplication, or the Identity Property of Multiplication. $$(-17)\cdot 1 = -17$$ Answer: $$(-4\cdot 2)\cdot 3 = -4(2\cdot 3)$$ Answer: $$(6\cdot 5)\cdot(-2) = (-2)\cdot(6\cdot 5)$$ Answer: $$(2)(3\cdot 6) = (2\cdot 3)(6)$$ Answer:	

Example	Notes
5. Multiply $\left(\dfrac{1}{4}\right)(-5)(-4)(3)$. Change the order of the terms to group simpler multiplications.	

Answer:

Helpful Hints

Changing the order of factors and grouping easier multiplications together may make the arithmetic of a problem simpler to perform.

Look for numbers that multiply to powers or multiples of ten, or fractions that will cancel, in each problem before you multiply.

Multiplying a number by zero gives us a product of zero. $a \cdot 0 = 0$
Multiplying a number by one gives us the same number. $1 \cdot a = a$

Concept Check

1. What makes this multiplication easy to simplify? $-9\left(6\dfrac{3}{4}\right)(-5.023)(0)\left(-\dfrac{5}{8}\right)(20,000)$

Practice

Multiply. Change the order of the terms to group simpler multiplications.

2. $(-9 \cdot 2) \cdot (5 \cdot 6)$

3. $9(-4)\left(\dfrac{2}{3}\right)(-5)$

Which property of multiplication is shown by each equation?

4. $(-9 \cdot 3) \cdot (-12) = (-12) \cdot (-9 \cdot 3)$

5. $(-7 \cdot 2) \cdot 5 = -7(2 \cdot 5)$

Variables and Expressions
Topic 4.1 Introduction to Expressions

Vocabulary

variable   •   term   •   algebraic expression   •   like terms   •   equation

1. A(n) _____ is a letter or symbol that is used to represent an unknown quantity.

2. A(n)_____ is any number, variable, or product of numbers and/or variables.

Step-by-Step Video Notes
Watch the Step-by-Step Video lesson and complete the examples below.

Example	Notes
1–5. Identify the terms in each expression.	

$3x$

Determine the number of terms.

Write the terms.

$2x + 1$

Determine the number of terms.

Write the terms.

$-7xy^2 + 3z - 2$

Determine the number of terms.

Write the terms.

60

Determine the number of terms.

Write the terms.

$-6x - 4z$

Determine the number of terms.

Write the terms.

Example	Notes

6–10.   Identify the like terms in each expression.

$3x + 4y - 7x + 5z$

$2x^2 + 3x$

$-4x + 5 - 2x - 7$

$2x^2 + 3x - 1 + 6x^2 - 9x + 8$

$8a - 7b + 3b + 2b + a$

Helpful Hints

An algebraic expression is a combination of numbers and variables, operation symbols, and grouping symbols.

Terms are the parts of an algebraic expression separated by a plus sign or a minus sign. The sign in front of the term is considered part of the term.

Like terms are terms that have the same variables raised to the same powers.

Concept Check

1. How many terms are in the algebraic expression $-6xy + 11xy^2 + 3.5$?

Practice

Identify the terms in each expression.

2. $4q + u$

Identify the like terms in each expression.

4. $7x + 4y - 3x + 9y - 5$

3. $-7ab^2 + 8c - 11$

5. $1.3x^3y + 5.6x + 3.2x^3y - 6.1x$

Variables and Expressions
Topic 4.2 Evaluating Algebraic Expressions

Vocabulary
value   •   evaluate   •   order of operations   •   variable

1. When you _____ an expression for a given value, you are finding the
    value of the expression for a given value of the variables.

Step-by-Step Video Notes
Watch the Step-by-Step Video lesson and complete the examples below.

Example	Notes
1. Evaluate $3x+6$ for $x=4$. Substitute the given value for the variable. $3(4)+6$ Simplify. Remember to follow the rules for order of operations. $\boxed{}+6$ Answer:	
2. Evaluate $5x^2$ for $x=6$. Substitute the given value for the variable. $5\left(\boxed{}\right)^2$ Simplify. Remember to follow the rules for order of operations. Answer:	

Example	Notes
3. Evaluate $3y^2 - y$ for $y = -7$. Answer:	
5. The perimeter of a rectangle can be found by the expression $2(l + w)$, where l is the length and w is the width. Find the perimeter of this rectangle if $l = 5.4$ and $w = 8$. Answer:	

Helpful Hints

Remember to use the Order of Operations (PEMDAS) when you're evaluating and simplifying algebraic expressions.

Many geometric measurements are found by evaluating algebraic expressions (called formulas) at given values.

Be sure to consider the sign of the value being substituted when evaluating algebraic expressions.

Concept Check

1. Explain how you would evaluate $4x^2 - 2x + 4$ for $x = 1.5$.

Practice

Evaluate for $x = 8$.

2. $-3x + 14$

Evaluate for the given values of l and w.

4. $2(l + w)$ for $l = 4$ and $w = 6.2$

3. $x^2 - 8x - 1$

5. $2l + 2w$ for $l = 17$ and $w = 16$

Name: _____   Date: _____
Instructor: _____   Section: _____

Variables and Expressions
Topic 4.3 Simplifying Expressions

Vocabulary
simplifying   •   like terms   •   algebraic expressions

1. You can simplify an expression by combining any _____.

Step-by-Step Video Notes
Watch the Step-by-Step Video lesson and complete the examples below.

Example	Notes
1–4. Combine like terms.	
$13x - 2x = \boxed{}\, x$	
$-4a + 3b + 9a = \boxed{}\, a + \boxed{}$	
$2x^3 + 9x^2 - x + 7$	
$\dfrac{4}{3}m + \dfrac{2}{3}m - m$	

5. Combine like terms.

$2a + 32b - 25c - 12b + 15a + 13c$

Rearrange the terms to group the like terms together.

Combine the like terms.

Answer:

Example	Notes
6. Combine like terms. $15x^3 + 2 - 8x^2 + x^3 - 9x + 13x^2 - 3$ Rearrange the terms to group the like terms together. Combine the like terms. Answer:	
7. Combine like terms. $16x^2y - 3xy^2 + 5xy - 2x^2y - 4xy^2$ Answer:	

Helpful Hints

If an expression has several terms, you might find it helpful to rearrange the terms before simplifying so that the like terms are together.

Remember that the sign in front of the term is part of the term. When you rearrange the terms, the sign in front of the term stays with the term.

The order in which you write the terms in the answer does not matter. However, it's customary to write the term with the highest exponent first, and number terms last.

Concept Check

1. Can you combine $2x^2y^3$ and $-5x^3y^2$? Explain.

Practice

Combine like terms.

2. $-5x + 24x$

3. $5x^3 + 3x^2 - 7x + 4$

4. $5x^3 + 3x^2 - 7x^3 + 4 - 2x^2$

5. $4 - 3y + 7y^2 + 9 - 5y^2 - 4 + 8y$

Variables and Expressions
Topic 4.4 Simplifying Expressions with Parentheses

Vocabulary

parentheses   •   grouping symbols   •   Distributive Property

1. If an algebraic expression contains parentheses and it is not possible to simplify what is inside the parentheses first, then you can apply the _____ to remove the parentheses.

Step-by-Step Video Notes
Watch the Step-by-Step Video lesson and complete the examples below.

Example	Notes
1. Simplify $5+4(a+b)$. Use the Distributive Property to remove the parentheses. $5+4a+\boxed{}$ Combine like terms. Answer:	
2. Simplify $-(4x-3y)$. Remove the parentheses by multiplying what is inside the parentheses by -1. Answer:	
6. Simplify $14x-2\left[3x+3(5)\right]$. Remove innermost parentheses first. Use the Distributive Property to remove the brackets. Then combine like terms. Answer:	

Example	Notes
7. Simplify. $$\dfrac{7-(4-x)}{3+2(x-5)}$$ Answer:	

Helpful Hints

A negative sign in front of parentheses means the opposite of what is inside the parentheses. Remove the parentheses by multiplying what is inside the parentheses by -1.

Work "inside out." Start with the innermost grouping symbols, and remove each set of grouping symbols in turn working from the inside to the outside.

When simplifying expressions within a fraction, simplify the expressions above and below the fraction bar first. Then simplify the fraction, if possible.

Concept Check

1. Explain how you would simplify $2\left[-4(9x-7)\right]$.

Practice
Simplify.

2. $8(x+y)+4(m+n)$

4. $5x+4\left[2x+6(3x-1)\right]$

3. $-(8a+3y-2)$

5. $\dfrac{4(-3y+9)}{7(5x-6)}$

Variables and Expressions
Topic 4.5 Translating Words into Symbols

Vocabulary
difference   •   quotient   •   Commutative Property

1. The order in which you write a subtraction or division expression matters, because the _____ does not work for subtraction or division.

2. Phrases such as "a number decreased by 6", "6 fewer than a number" and "the _____ between a number and 6" all indicate subtraction.

Step-by-Step Video Notes
Watch the Step-by-Step Video lesson and complete the examples below.

Example	**Notes**
1–3. Translate into an algebraic expression.	
The sum of 8 and a number	
Triple a number	
75% of a number	
4–7. Translate into an algebraic expression.	
Five less than twelve	
Twelve less than five	
Fifty divided by one	
One divided by fifty	

Example	Notes
11–13. Translate into an algebraic expression. Use parentheses if necessary. Seven more than double a number x A number x is tripled and then increased by 8 One-half the sum of a number x and 3	

14. Use an expression to describe the measure of each angle.

Second angle

First angle   Third angle

The measure of the second angle of a triangle is double the measure of the first angle, and the third angle is 15° more than the measure of the second angle.

Helpful Hints
The order of the terms is important in subtraction and division expressions.

Use parentheses when writing expressions to be sure certain operations are performed first.

Concept Check
1. What are some words or phrases in an expression that indicate addition?

Practice
Translate into an algebraic expression.

2. A number n increased by seven

3. The quotient of 24 and some number y

4. Triple the difference of a and b

5. 100 less than the product of 3 and x

Introduction to Solving Linear Equations
Topic 5.1 Translating Words into Equations

Vocabulary
variable   •   equation   •   expression

1. A(n) _____ is a letter or symbol that is used to represent an unknown quantity.

2. A(n) _____ is a mathematical statement that two expressions are equal.

Step-by-Step Video Notes
Watch the Step-by-Step Video lesson and complete the examples below.

Example	Notes
1. Translate into an equation. Let n represent the number. One-third of a number is fourteen. Look for key words to help you translate the words into algebraic symbols and expressions. One-third of a number is fourteen. $$\frac{\square}{\square} \cdot \square = \square$$ The equation is:	
2. Translate into an equation. Do not solve. Five more than six times a number is three hundred five. The equation is:	

Example	Notes
3. Translate into an equation. Do not solve. The larger of two numbers is three more than twice the smaller number. The sum of the numbers is thirty-nine. Write an expression to represent each unknown quantity in terms of the chosen variable. Let s stand for the smaller number. An expression for the larger number is _____. The equation is:	
4. Translate into an equation. Do not solve. The annual snowfall in Juneau, Alaska is 105.8 inches. This is 20.2 inches less than three times the annual snowfall in Boston, Massachusetts. The equation is :	

Helpful Hints
When writing an equation from a word problem, write an expression to represent each unknown quantity in terms of the variable. Use a given relationship in the problem or an appropriate formula to write an equation.

Concept Check
1. What words or phrases are typically associated with multiplication? Division? Addition? Subtraction?

Practice
Translate into an equation. Do not solve.

2. A number n increased by three is nine.

4. Twice the difference of a and 7 is eight.

3. One-fourth of a number is 16.

5. 100 less than the product of 3 and x is 5.

Introduction to Solving Linear Equations
Topic 5.2 Linear Equations and Solutions

Vocabulary
equation • solution • variable • linear equation in one variable • exponent

1. A(n) _____ of an equation is the number(s) that, when substituted for the variable(s), makes the equation true.

2. A(n) _____ is an equation that can be written in the form $Ax + B = C$ where $A, B,$ and C are real numbers and $A \neq 0$.

Step-by-Step Video Notes
Watch the Step-by-Step Video lesson and complete the examples below.

Example	Notes
1. Is 2 a solution of the equation $3x - 1 = 5$? Substitute 2 for x. $3(\square) - 1 \overset{?}{=} 5$ Simplify each side of the equation. $\square \overset{?}{=} 5$ Answer:	
2 & 3. Is -1 a solution of the equation $2x + 6 = -1$? Is -3 a solution to $7x - 2 = 5$? Answer:	

Example	Notes
4–6. Determine if each equation is a linear equation. $2x + 3 = 1$ $2x = 5$ $6x^2 - 3 = 4$	

Helpful Hints

The variable in a linear equation cannot have an exponent greater than 1.

To determine if a given value is a solution of an equation, substitute the given value into the equation. Simplify each side, and if the result is a true statement, that value is a solution.

Concept Check

1. Why is the equation $x(x + 4) = 45$ not a linear equation?

Practice

Is 7 a solution of the equation?

2. $42 - 3x = 21$

Determine if each is a linear equation.

4. $5x + 4\frac{1}{3} = -9$

3. $31 = 4x + 5$

5. $11x - 18 = x^2$

Introduction to Solving Linear Equations
Topic 5.3 Using the Addition Property of Equality

Vocabulary
equivalent equations　　•　　solving the equation　　•　　Addition Property of Equality

1.　The process of finding the solution(s) of an equation is called _____.

2.　Equations that have exactly the same solutions are called _____.

Step-by-Step Video Notes
Watch the Step-by-Step Video lesson and complete the examples below.

Example	Notes
1.　Solve $x+16=20$ for x. Check your solution. Use the Addition Property to subtract 16 from both sides. Simplify. Check your solution. Solution:	
2.　Solve $-14=x-3$ for x. Check your solution. Add $\boxed{}$ to both sides of the equation. Simplify. Check your solution. Solution:	

Example	Notes
3. Solve $x - \dfrac{1}{2} = -\dfrac{5}{2}$ for x. Check your solution. Add $\dfrac{\square}{\square}$ to both sides of the equation. Solution:	
4. Solve $15 + 2 = 3 + x + 6$ for x. Check your solution. Solution:	

Helpful Hints

To solve an equation using the Addition Property, add or subtract the same number from both sides of the equation to get the variable x on a side of the equation by itself. If a number is being added to x, use subtraction. If a number is being subtracted, use addition.

Concept Check

1. Explain why the equations $23 - 5 = 4 + x + 3$ and $11 = x + 3$ are equivalent.

Practice

Solve for x. Check your solution.

2. $x + 19 = 28$

3. $-23 = x - 6$

4. $x - \dfrac{3}{5} = \dfrac{7}{5}$

5. $-25 + 12 = -8 + x + 16$

Introduction to Solving Linear Equations
Topic 5.4 Using the Multiplication Property of Equality

Vocabulary

Addition Property of Equality   •   reciprocal   •   Multiplication Property of Equality

1. The _____ states that if both sides of an equation are multiplied by the same non-zero number, the solution does not change.

2. Dividing by a number is the same as multiplying by its _____.

Step-by-Step Video Notes
Watch the Step-by-Step Video lesson and complete the examples below.

Example	Notes
1. Solve $\frac{x}{3} = -15$ for x. Check your solution. Use the Multiplication Property to multiply each side of the equation by 3. Simplify. Check your solution. Solution:	
2. Solve $3x = 21$ for x. Check your solution. Divide both sides by $\boxed{}$. Simplify. Check your solution. Solution:	

Example	Notes
3. Solve $20 = -4x$ for x. Check your solution. Solution:	
4. Solve $2x - 5x = -12$ for x. Check your solution. Simplify each side of the equation. Solution:	

Helpful Hints

To solve an equation using the Multiplication Property of Equality, multiply or divide both sides of the equation by the same number to get the variable x on a side of the equation by itself. If x is being multiplied by a number, use division. If x is being divided by a number, use multiplication.

Concept Check

1. By what number should you multiply each side of the equation $\frac{1}{6}x = -4$ to solve for x?

Practice

Solve for x. Check your solution.

2. $-32 = -4x$

3. $-0.4x = 2.8$

4. $\frac{x}{3} = -9$

5. $7x - 8x = 9$

Introduction to Solving Linear Equations
Topic 5.5 Using the Addition and Multiplication Properties Together

Vocabulary
Addition Property of Equality   •   variable term   •   Multiplication Property of Equality

1. To solve an equation of the form $ax + b = c$, we must use both the _____

   and the _____ together.

Step-by-Step Video Notes
Watch the Step-by-Step Video lesson and complete the examples below.

Example	**Notes**
1. Solve $5x + 3 = 18$ for x to determine how many goals Jenny scored, and then check your solution. Use the Addition Property to subtract 3 from both sides. Use the Multiplication Property to divide both sides by 5. Solution:	
2. Solve $-\dfrac{1}{2}x + 10 = 16$ for x. Use the Addition Property to subtract ☐ from both sides. Use the Multiplication Property to multiply both sides by ☐. Solution:	

Example	Notes
3. Solve $6x - 8 = -2$ for x. Solution:	
4. Solve $4 = -7 + 8x$ for x. Solution:	

Helpful Hints

To evaluate an equation of the form $ax + b = c$, the order of operations tells us to multiply before adding. When trying to solve the equation for x, we must undo this. That is, we must add (or subtract) first, and then multiply (or divide).

Concept Check

1. Which operation would you undo first to solve the equation $-2x + 8 = -14$ for x?

Practice

Solve for x.

2. $7x + 3 = 45$

4. $\dfrac{1}{6}x + 4 = -8$

3. $-22 = 3x - 7$

5. $4x - 13.2 = 14.8$

Solving More Linear Equations and Inequalities
Topic 6.1 Solving Equations with Variables on Both Sides

Vocabulary
solution   •   combining like terms   •   solving the equation

1. The process of finding the solution(s) of an equation is called _____.

Step-by-Step Video Notes
Watch the Step-by-Step Video lesson and complete the examples below.

Example	Notes
1. Solve $9x = 6x + 15$ for x. The goal is to get the variable alone on one side of the equation and numbers on the other side. Subtract $6x$ from both sides. Solution:	
2. Solve $9x + 4 = 7x - 2$ for x. The goal is to get the variable alone on one side of the equation and numbers on the other side. Subtract $\boxed{} x$ from both sides. Subtract $\boxed{}$ from both sides. Solution:	

Example	Notes
3. Solve $5x + 26 - 6 = 9x + 12x$ for x.	
Simplify each side.	
Get the variable terms on one side.	
Solution:	
4. Solve $-x + 8 - x = 3x + 10 - 3$ for x.	
Solution:	

Helpful Hints

Sometimes variable terms and number terms appear on both sides of the equation. If it is necessary, simplify one or both sides of the equation by combining like terms that are on the same side of the equation. Then get the variable terms on one side of the equation and the number terms on the other side.

Concept Check

1. Which term would you add to both sides of the equation $-2x - 8 = 14 - 5x$ so that the variable terms are on one side of the equation with a positive coefficient?

Practice

Solve for x.

2. $4x + 6 = 8x$

4. $7 - 3x + 2 = -9 + 4x - 10$

3. $9x - 22 = 3x - 4$

5. $-12x + 2 + 7x = 1 - 8x + 16$

Solving More Linear Equations and Inequalities
Topic 6.2 Solving Equations with Parentheses

Vocabulary
Distributive Property   •   parentheses   •   equation

1. In order to solve an equation with parentheses, simplify by using the _____
   to remove the parentheses.

Step-by-Step Video Notes
Watch the Step-by-Step Video lesson and complete the examples below.

Example	Notes
1. Solve $2(x+5) = -12$ for x. Simplify each side. $2\Box + \Box = -12$ Notice there is only one variable term. Get the number terms on the other side. Get the variable alone on one side. Solution:	
3. Solve $5(x+1) - 3(x-3) = 17$ for x. Simplify each side. $\Box x + \Box - \Box x + \Box = 17$ Get the variable terms on one side. Get the number terms on the other side. Get the variable alone on one side. Solution:	

Example	Notes
5. Solve $3(0.5x - 4.2) = 0.6(x - 12)$ for x. Solution:	
6. Solve $2(18x - 5) + 2 = 24x - 3(12x + 8)$ for x. Solution:	

Helpful Hints

Recall that the Distributive Property states that for all real numbers $a, b,$ and c,

$a(b + c) = ab + ac$. Sometimes an equation has multiple sets of parentheses. If this is the case, apply the Distributive Property as many times as is necessary to remove all sets of parentheses. Also, remember that parentheses can appear inside other parentheses.

Concept Check

1. Can you solve the equation $3(x - 7) = 12$ without using the Distributive Property?

Practice

Solve for x.

2. $4(x + 6) = -8$

3. $7(x - 2) - 6 = x + 4$

4. $7(-3x + 2) = -8(4x - 10)$

5. $0.3x - 2(x - 1.2) = -0.7(x - 3) - 3.7$

Solving More Linear Equations and Inequalities
Topic 6.3 Solving Equations with Fractions

Vocabulary
Distributive Property   •   least common denominator   •   equivalent equation

1. To make the process of solving equations with fractions easier, multiply both sides of the equation by the _____ of all the fractions contained in the equation.

Step-by-Step Video Notes
Watch the Step-by-Step Video lesson and complete the examples below.

Example	Notes
1. Solve $\dfrac{1}{4}x - \dfrac{2}{3} = \dfrac{5}{12}x$ for x. Find the LCD of the fractions, then multiply both sides of the equation by the LCD. $\Box\left(\dfrac{1}{4}x - \dfrac{2}{3}\right) = \Box\left(\dfrac{5}{12}x\right)$ Use the Distributive Property. Solution:	
2. Solve $\dfrac{x}{3} + 3 = \dfrac{x}{5} - \dfrac{1}{3}$ for x. Find the LCD of the fractions, then multiply both sides of the equation by the LCD. $\Box\left(\dfrac{x}{3} + 3\right) = \Box\left(\dfrac{x}{5} - \dfrac{1}{3}\right)$ Solution:	

Example	Notes
3. Solve $\dfrac{x+5}{7} = \dfrac{x}{4} + \dfrac{1}{2}$ for x. Find the LCD of the fractions, then multiply both sides of the equation by the LCD. Solution:	
4. Solve $0.6x - 1.3 = 4.1$ for x. Solution:	

Helpful Hints

You can also solve an equation containing decimals in a similar way to the fraction equations. You can multiply both sides of the equation by an appropriate value to eliminate the decimal numbers and work only with integer coefficients. If the decimals are tenths, multiply by 10, if the decimals are hundredths or less, then multiply by 100, etc.

Concept Check

1. By what number would you multiply each term in the equation $0.03x - .42 = 1.2$ to work with only integer coefficients?

Practice

Solve for x.

2. $\dfrac{1}{2}x - \dfrac{2}{3} = \dfrac{5}{6}$

4. $\dfrac{x+6}{12} = \dfrac{x}{6} + \dfrac{3}{4}$

3. $\dfrac{7}{8}x - \dfrac{5}{2} = \dfrac{3}{4}x$

5. $3.6 = 4(0.6x - 0.3)$

Solving More Linear Equations and Inequalities
Topic 6.4 Solving a Variety of Equations

Vocabulary
identity   •   infinite number of solutions   •   no solution
contradiction   •   solving an equation

1. An equation has _____ if there is no value of x that makes the equation true.

2. An equation has an _____ if the equation is always true, no matter the value of x.

Step-by-Step Video Notes
Watch the Step-by-Step Video lesson and complete the examples below.

Example	Notes
1. Solve $3(6x-4)=4(3x+9)$ for x. Remove the parentheses using the Distributive Property. Solution:	
2. Solve $2(3x+1)=5(x-2)+3$ for x. Solution:	

Example	Notes
3. Solve $\frac{1}{3}(x-2)=\frac{1}{5}(x+4)+2$ for x. Solution:	
4. Solve $5(x+3)=2x-8+3x$ for x. Solution:	

Helpful Hints

Checking the solution is arguably the most important step of solving an equation. There are situations where possible solutions found may not actually be solutions.

It is important to follow the order of operations when solving equations.

Concept Check
1. Why is an equation such as $2x+9+x=4+3x+5$ called an identity?

Practice
Solve for x.

2. $\frac{1}{4}(4x-12)=\frac{1}{5}(10x+5)$

4. $-6+4(x-3)=11x-5-7x$

3. $\frac{1}{3}(x+6)=\frac{1}{6}(x-3)+\frac{2}{3}$

5. $7(x+3)-6=24-4x-9+11x$

Solving More Linear Equations and Inequalities
Topic 6.5 Solving Equations and Formulas for a Variable

Vocabulary
formula   •   Distributive Property   •   least common denominator

1. A _____ is an equation in which variables are used to describe a relationship.

Step-by-Step Video Notes
Watch the Step-by-Step Video lesson and complete the examples below.

Example	Notes
1. Solve $5x + 2 = 17$ and $ax + b = c$ for x. Identify the variable in both equations. Notice that these equations are of the same form, except that every number in the first equation is a variable in the second equation. $$5x + 2 = 17 \qquad\qquad ax + b = c$$ $$5x + 2 - \square = 17 - \square \qquad ax + b - \square = c - \square$$ $$\frac{5x}{\square} = \frac{\square}{\square} \qquad\qquad \frac{ax}{\square} = \frac{\square}{\square}$$ $$x = \square \qquad\qquad\qquad x = \frac{\square}{\square}$$ Solution:	
2. Solve $d = rt$ for t. Divide both sides of the equation by r. Solution:	

Example	Notes
4. Solve for the specified variable. $a = \dfrac{v}{t}, \; v$ Solution:	
7. Solve for the specified variable. $5x + 3y = 6,$ solve for y Solution:	

Helpful Hints

To solve a formula or an equation for a specified variable, use the same steps for solving a linear equation except treat the specified variable as the only variable in the equation and treat the other variables as if they were numbers.

Sometimes if there are two variables in an equation, you may be asked to solve for one variable *in terms of* the other. For example, "solve $2x - 3y = 6$ for y" indicates that you are to find y in terms of x.

Concept Check

1. What would you divide by to solve $A = bh$ for h?

Practice

Solve for x.

2. $2x + 8y = 12$

Solve for the specified variable.

4. $A = \dfrac{1}{2}bh$, solve for h.

3. $y = mx + b$

5. $8x - 3y = 12$, solve for y.

Solving More Linear Equations and Inequalities
Topic 6.6 Solving and Graphing Inequalities

Vocabulary

inequality   •   linear inequality in one variable   •   solution of an inequality
graph of an inequality   •   solve an inequality   •   non-solutions

1. The _____ is a picture that represents all of the solutions of the inequality.

2. A(n) _____ is a statement that shows the relationship between any two
   real numbers that are not equal.

Step-by-Step Video Notes

Watch the Step-by-Step Video lesson and complete the examples below.

Example	Notes
1 & 2. Graph each inequality on a number line.	

$x > 3$

Use a(n) _____ circle at the boundary
point $x = 3$, because 3 _____ a solution.

Shade all numbers to the _____ side of
the boundary point.

$x \leq -1$

Use a(n) _____ circle at the boundary
point $x = -1$, because −1 _____ a
solution.

Shade all numbers to the _____ side of
the boundary point.

Example	Notes
3. Solve and graph the inequality $5x + 2 < 12$. Subtract 2 from both sides of the inequality. Divide both sides of the inequality by 5. Graph the inequality.	

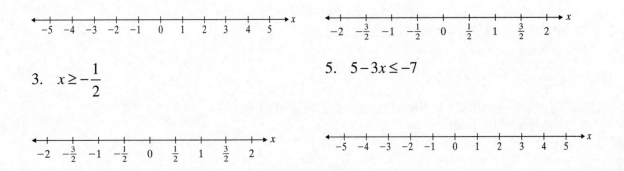

4. Solve and graph the inequality $5 - 4x \geq -7$.

Helpful Hints

It is important to decide if you need an open circle or a closed circle. Remember, for the endpoint, use an open circle for $<$ or $>$ and a closed circle for $\leq$ or $\geq$.

Use the same procedure to solve an inequality that is used to solve an equation, *except* the direction of an inequality must be *reversed* if you *multiply* or *divide* both sides of the inequality *by a negative* number.

Concept Check

1. Would you use an open or closed circle for the boundary point to graph $2x + 3 \leq 7$?

Practice

Graph the inequality on a number line.

2. $x < 2$

Solve and graph the inequality.

4. $8 + 4x > 6$

3. $x \geq -\dfrac{1}{2}$

5. $5 - 3x \leq -7$

Introduction to Graphing Linear Equations
Topic 7.1 The Rectangular Coordinate System

Vocabulary
rectangular coordinate system   •   *x*-coordinate   •   *y*-coordinate   •   ordered pair

1. A _____ is made up of a horizontal number line and a vertical number line that intersect to form a right angle. The point where these number lines meet is the origin.

Step-by-Step Video Notes
Watch the Step-by-Step Video lesson and complete the examples below.

Example	Notes
1. Plot the points $(4,2)$, $(3,-2)$, $(-3,3)$, and $(-1,-4)$. Label them A, B, C, and D, respectively.	
2. Plot the points on the graph. Identify which quadrant each point lies in. Label them A, B, C, and D, respectively. a. $(4,-5)$ b. $(-5,4)$ c. $(3,0)$ d. $(2,2)$	

Example	Notes
3. Find the coordinates of the indicated points. Write each point as an ordered pair. Identify which quadrant each point lies in. 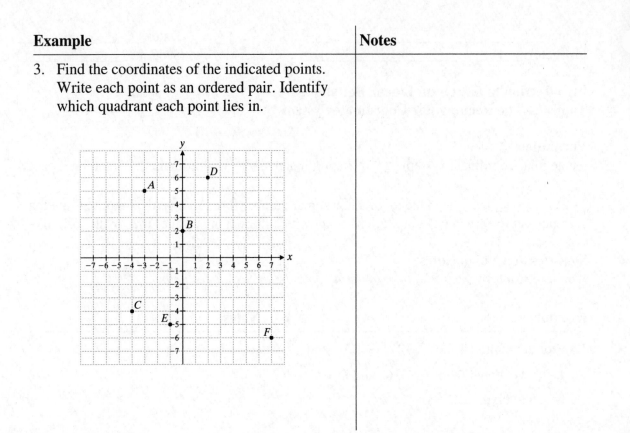	

Helpful Hints

The origin is represented by the ordered pair $(0,0)$, and has both x- and y-coordinates of 0.

The quadrants are numbered I, II, III, and IV, starting at the top right and going counter-clockwise. Points that lie on an axis are not considered to be in any quadrant.

Concept Check
1. Give an example of a point in Quadrant IV. Give an example of a point not in a quadrant.

Practice
Plot the given points on the graph. Identify which quadrant each point lies in.

2. $A(-3,4)$ 3. $B(4,-5)$ 4. $C(0,-2)$ 5. $D(4,1)$

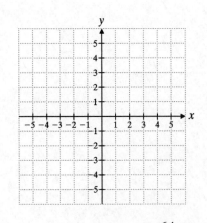

Introduction to Graphing Linear Equations
Topic 7.2 Graphing Linear Equations by Plotting Points

Vocabulary
linear equations in two variables • solution to an equation • ordered pair

1. A _____ is an equation that can be written in the form $Ax + By = C$
   where A, B, and C are real numbers, but A and B are not both zero.

Step-by-Step Video Notes
Watch the Step-by-Step Video lesson and complete the examples below.

Example	Notes
1. Determine whether $(3,5)$ is a solution to the equation $3x + 2y = 19$. Substitute the x- and y-coordinates into the linear equation. Check for a true statement. $3\left(\boxed{}\right) + 2\left(\boxed{}\right) \overset{?}{=} 19$ $\boxed{} \overset{?}{=} 19$ Answer:	
4. Find three solutions to $2x + y = 13$. Substitute a value for one of the variables. Solve the equation for the other variable. Repeat for the other two solutions. Write the ordered pairs. Answer:	

Example	Notes
6. Find three solutions to $x + y = -4$. Make a table of values to keep your ordered pairs organized. Answer:	
7. Graph the equation $y = 2x + 1$. Make a table of values to find three ordered pair solutions. Plot the ordered pairs on a graph. Draw a line through the points. 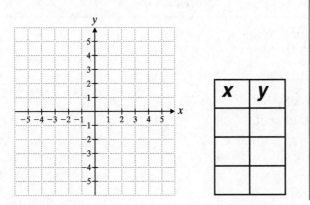	

Helpful Hints

To graph a linear equation in two variables, find at least 3 ordered pairs which are solutions of the equation. Plot the ordered pairs, and then draw a line through the points.

Concept Check

1. How many ordered pairs are solutions to the equation $37.5x - 19.2y = 4.8$?

Practice

Is the given point a solution of $2x + 8y = 20$?

2. $(2, 2)$

3. $(-6, 4)$

Find three solutions to the given equation.

4. $2x - y = 7$

5. $y = -4x + 1$

Introduction to Graphing Linear Equations
Topic 7.3 Graphing Linear Equations Using Intercepts

Vocabulary

intercept   • x-intercept   • y-intercept   •   origin

1. The point at which a line crosses an axis is called a(n) _____.

2. The _____ is an ordered pair with the coordinates $(a, 0)$, where a is a real number.

Step-by-Step Video Notes

Watch the Step-by-Step Video lesson and complete the examples below.

Example	Notes
1. Find the x-intercept and y-intercept of $3x - 6y = 12$. Find the x-intercept by letting $y = 0$ and solving for x. Find the y-intercept by letting $x = 0$ and solving for y. Answer:	
2. Find the x-intercept and y-intercept of $y = \dfrac{4}{5}x$. Find the x-intercept by letting $y = 0$ and solving for x. Find the y-intercept by letting $x = 0$ and solving for y. Answer:	

Example	Notes

3. Graph $2x - y = 4$ using the intercepts.

Make a table of values to find the x-intercept, y-intercept, and another value.

Plot the points on the graph and draw a line through the points.

x	y
	0
0	

Helpful Hints

Graphing an equation using the intercepts is exactly the same process as graphing lines by plotting points. In this method, you use two specific points, (the x-intercept and the y-intercept) then plot one more ordered pair and draw a line through the points.

Concept Check

1. Give an example of a linear equation in two variables where both intercepts are the origin, $(0,0)$.

Practice

Find the x- and y-intercepts of each equation.

2. $5x - 4y = 20$

4. $2 - y = 3x + 2$

3. $y = \dfrac{1}{2}x + 3$

5. $y = -4x + 1$

Introduction to Graphing Linear Equations
Topic 7.4 Graphing Linear Equations of the Form $x = a$, $y = b$, $y = mx$

Vocabulary
origin   •   horizontal line   •   vertical line

1. If an equation is of the form $y = b$, where b is some real number, then the graph of the equation is a _____.

2. If an equation is of the form $x = a$, where a is some real number, then the graph of the equation is a _____.

Step-by-Step Video Notes
Watch the Step-by-Step Video lesson and complete the examples below.

Example	Notes
1. Graph $6x - 2y = 0$ using three points. Make a table of values. Then, plot the points on a graph and draw a line through the points. 	
2. Graph the equation $y = 4$. 	

Example	Notes

5. Graph $3x + 7 = -5$

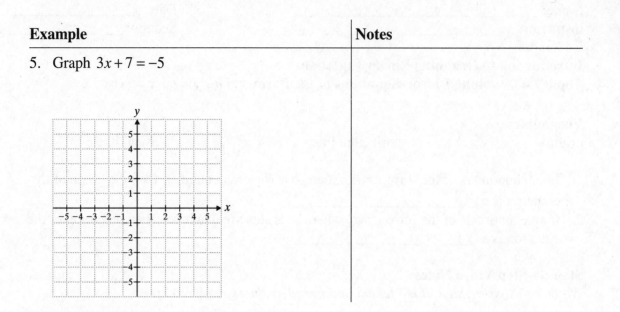

Helpful Hints
The graph of the vertical line has no y-intercept, unless its equation is $x = 0$.

Similarly, the graph of a horizontal line has no x-intercept, unless its equation is $y = 0$.

Any equation of the form $y = mx$ is neither vertical nor horizontal, and its x- and y-intercepts are the origin, $(0, 0)$.

Concept Check
1. The graph of what equation would include all of the points on the x-axis? What are its intercepts?

Practice
Graph each equation.

2. $2x - 4y = 0$ 3. $y - 1 = 2$ 4. $x = -2$

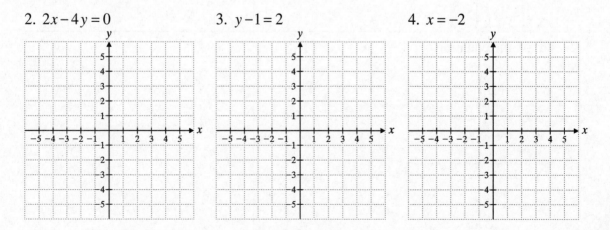

Slope and the Equation of a Line
Topic 8.1 The Slope of a Line

Vocabulary

rise   •   run   •   slope   •   ordered pair

1. The _____ of a line is the change in horizontal position, or the difference between the x-coordinates of two points on the line.

2. The _____ of a line is the rate of change between any two ordered pair solutions to a linear equation.

Step-by-Step Video Notes
Watch the Step-by-Step Video lesson and complete the examples below.

Example	Notes
1. Find the slope. Answer:	
3. Find the slope. Answer:	

Example	Notes
5. Find the slope of the line which contains the points $(1,3)$ and $(5,11)$. Use the slope formula $m = \dfrac{y_2 - y_1}{x_2 - x_1}$. $$m = \frac{\boxed{} - \boxed{}}{\boxed{} - \boxed{}}$$ Answer:	

Helpful Hints

The formula for slope is $\text{Slope} = \dfrac{\text{change in } y\text{-coordinates}}{\text{change in } x\text{-coordinates}} = \dfrac{y_2 - y_1}{x_2 - x_1}$ where $x_1 \neq x_2$.

The variable m is typically used to represent slope.

A line with a positive slope goes up from left to right. A line with a negative slope goes down from left to right. The slope of a horizontal line is zero. The slope of a vertical line is undefined; it can also be said to have no slope. No slope is not the same as zero slope.

Concept Check

1. A line's run is positive and its rise is negative. Is the slope of this line positive or negative?

Practice

Find the slopes of the lines containing the given points.

Find the slopes of the lines.

2. $(2,8)$ and $(0,0)$

4. $y = 4$

3. $(3,1)$ and $(-5,5)$

5. $x = -2.5$

Slope and the Equation of a Line
Topic 8.2 Slope-Intercept Form

Vocabulary
slope-intercept form of a linear equation   • y-intercept   •   Standard Form

1. The _____ that has a slope m and a y-intercept $(0, b)$
    is given by the formula $y = mx + b$.

Step-by-Step Video Notes
Watch the Step-by-Step Video lesson and complete the examples below.

Example	Notes
1. Find the slope and y-intercept of $y = \dfrac{2}{3}x + 5$. Find m. $\dfrac{\Box}{\Box}$ Find b. $\Box$ Answer:	
3. Find the slope and y-intercept of $4x - 3y = 12$. Rewrite the equation in slope-intercept form. Answer:	
6. Find the equation of a line with a slope of 2 and a y-intercept of $\left(0, \dfrac{4}{3}\right)$. Substitute the values for slope and y-intercept into the slope-intercept form. Answer:	

Example	Notes
8. Write the equation of the line shown in the graph. Answer:	

Helpful Hints

You can write an equation in slope-intercept form by solving it for y.

Horizontal lines of the form $y = b$ are simplified from $y = 0x + b$, the slope-intercept form of a line with zero slope. Vertical lines of the form $x = a$ cannot be written in slope-intercept form since they have an undefined slope.

Concept Check

1. How do you know the graph of the equation $y = -3x + 5$ passes through the point $(0, 5)$?

Practice

Find the slope and y-intercept of the given equation.

2. $x - 2y = 6$

Write the equation of a line with the given slope and y-intercept.

4. $m = 0.5$, $b = -2$

3. $y = 3x - 4$

5. $m = -\dfrac{3}{4}$, $b = 6$

Slope and the Equation of a Line
Topic 8.3 Graphing Lines Using Slope and *y*-Intercept

Vocabulary

slope   •   *y*-intercept   •   rise   •   run

1. The _____ of a line can be described as "the rise over the run."

Step-by-Step Video Notes
Watch the Step-by-Step Video lesson and complete the examples below.

Example	Notes
1. Find the *y*-intercept of the graph.	

$b = \boxed{}$

Answer:

3. Find the slope and the *y*-intercept of the line.

Answer:

Example	Notes

4. Graph $y = \dfrac{1}{2}x - 3$.

The y-intercept is $\boxed{}$.

The slope is $\dfrac{\boxed{}}{\boxed{}}$.

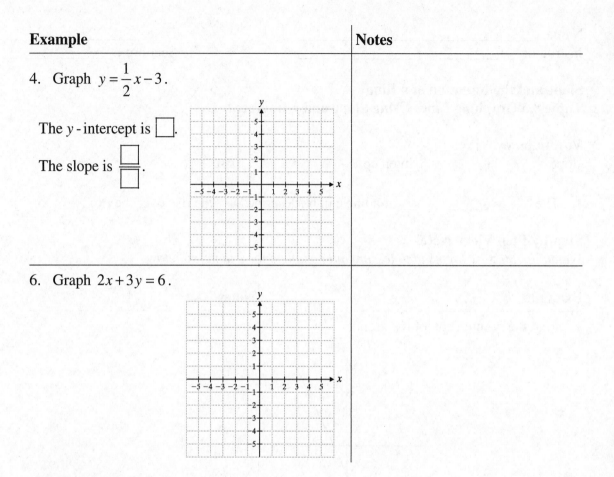

6. Graph $2x + 3y = 6$.

Helpful Hints

If the slope is an integer, its denominator is 1. If x has no coefficient in slope-intercept form, the slope is 1. If there is no constant b in slope-intercept form, the y-intercept is 0.

Concept Check

1. A line has a y-intercept of -2. Explain how to plot another point if the slope is $-\dfrac{2}{3}$.

Practice
Graph each equation.

2. $y = x + 1$

3. $y = \dfrac{1}{3}x - 1$

4. $y = -2x + 4$

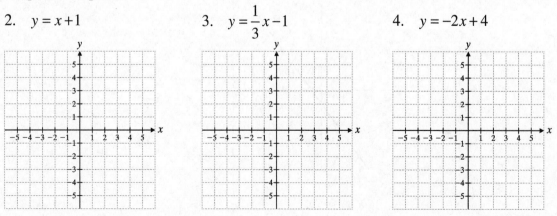

Slope and the Equation of a Line
Topic 8.4 Writing Equations of Lines Using a Point and Slope

Vocabulary
Slope-intercept form of a linear equation   •   point-slope form   •   standard form

1. The _____ of an equation of a line whose slope is m and passes through the point (x_1, y_1) is given by $y - y_1 = m(x - x_1)$.

Step-by-Step Video Notes
Watch the Step-by-Step Video lesson and complete the examples below.

Example	Notes
1. Use the point-slope form to write the equation of the line with a slope of 2 and passes through $(2,1)$. Write the point-slope equation. Substitute the given values of x_1, y_1, and m into the equation. Simplify. Answer:	
2. Write the equation of the line with a slope of -3 and passes through $(3,9)$. Write your answer in slope-intercept form. Write the point-slope equation. Solve the equation for y. Answer:	

Example	Notes
3. Write the equation of the line with a slope of $\frac{1}{7}$ and passes through $(14,-5)$. Write your answer in slope-intercept form. Answer:	
4. Use the point-slope form to write the equation of the line with a slope of 0 and passes through $(-16,-9)$. Answer:	

Helpful Hints

Point-slope form tells the slope of the line and the coordinates of a point on the line (not necessarily the y-intercept).

Linear equations are usually not left in point-slope form. Most of the time, they will be expressed in slope-intercept form.

Concept Check

1. What is the slope of the graph of the equation $y-32.7=1.5(x-41.3)$?

Practice

Write the equation of the line with the given slope and passes through the given point. Leave your answer in point-slope form.

2. slope of -4 and passes through $(-6,5)$

Write the equation of the line with the given slope and passes through the given point. Write your answer in slope-intercept form.

4. slope of 3 and passes through $(5,7)$

3. slope of $\frac{3}{7}$ and passes through $\left(2,-\frac{8}{9}\right)$

5. slope of $\frac{1}{2}$ and passes through $(-2,-1)$

Slope and the Equation of a Line
Topic 8.5 Writing Equations of Lines Using Two Points

Vocabulary
slope-intercept form   •   point-slope form   •   intercept

1. To write the equation of a line when given two points, use the points to find the slope, then pick one of the points and write the equation in _____.

Step-by-Step Video Notes
Watch the Step-by-Step Video lesson and complete the examples below.

Example	Notes
1. Write the equation in slope-intercept form of the line which passes through $(2,5)$ and $(6,3)$. Find the slope m. Substitute the values of x_1, y_1, and m into the point-slope form. Solve for y to write the answer in slope-intercept form. Answer:	
3. Write the equation of the line which passes through $(-5,5)$ and $(0,-6)$. Write your answer in slope-intercept form. Answer:	

Example	Notes
4. Write the equation of the line which passes through $(-8, 4)$ and $(-5, 0)$. Write your answer in slope-intercept form.	
Answer:	

5. Write the equation of the line on the graph. Write your answer in slope-intercept form.

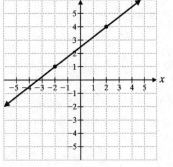

Answer:

Helpful Hints

Sometimes when given two points to find a line, one of the points is the y-intercept. If this is the case, you can find the slope and then use that point to write the equation in slope-intercept form directly rather than using point-slope form first.

Concept Check

1. Find the slope-intercept form of the line through the points $(-6, 5)$ and $(0, 3)$ without using point-slope form.

Practice

Write the equation of the line that passes through the given points.

2. $(-2, -5)$ and $(8, 5)$

3. $(-8, 13)$ and $(-6, 5)$

4. $(-11, -4)$ and $(9, 8)$

5. $(-2, -2)$ and $(10, 4)$

Introduction to Functions
Topic 9.1 Relations and Functions

Vocabulary
relation   •   domain   •   range   •   function   •   ordered pair

1. The second coordinates, or y values, in all of the ordered pairs of a relation make up the _____ of the relation.

2. A _____ is a relation for which every x value in the domain has one and only one y value.

Step-by-Step Video Notes
Watch the Step-by-Step Video lesson and complete the examples below.

Example	Notes
1. State the domain and range of the relation. $\{(5,7),(9,11),(10,7),(12,14)\}$ State the domain. State the range. Answer:	
2 & 3. Determine whether each relation is a function. $\{(3,9),(4,16),(5,9),(6,36)\}$ $\{(7,8),(9,10),(12,13),(7,14)\}$.	

Example	Notes

4–6. Determine whether each relation is a function.

$y = 3x - 5$

x			
y			

$y = |x|$

x			
y			

$y^2 = x$

x			
y			

Helpful Hints

A relation is any set of ordered pairs. Some relations cannot be expressed by an equation.

The first coordinates, or x values, in all of the ordered pairs of a relation make up the domain of the relation.

Concept Check

1. If you switch the order of the ordered pairs in the relation $\{(-2,4),(-1,1),(1,1),(2,4)\}$, will it still be a function? Explain.

Practice

State the domain and range of the relation.

2. $(8,5)\ \{(3,9),(9,3),(4,6),(6,4)\}$

3. $y = x^2 - 1$

Determine whether each relation is a function

4. $x = |y|$

5. $y = x^2 + 4$

Introduction to Functions
Topic 9.2 The Vertical Line Test

Vocabulary
vertical line test   • x-axis   •   ordered pairs

1. The _____ states that if a vertical line can pass along the x-axis
   and cross the graph in at most one place, then the graph represents a function.

Step-by-Step Video Notes
Watch the Step-by-Step Video lesson and complete the examples below.

Example	Notes
1. Determine whether the following is the graph of a function. Answer:	
2. Determine whether the following is the graph of a function. Answer:	

Example	Notes

3. Determine whether the following is the graph of a function.

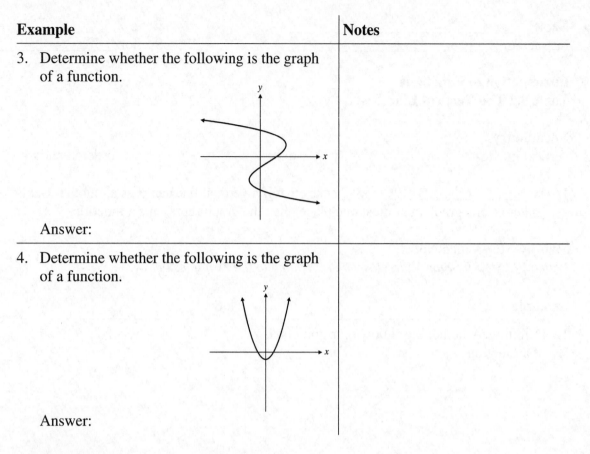

Answer:

4. Determine whether the following is the graph of a function.

Answer:

Helpful Hints
A function cannot have two different ordered pairs with the same first coordinate. That is, each value of x must have one and only one value of y.

Concept Check
1. Explain why $y = -2$ is a function, but why $x = -2$ is not a function.

Practice
Determine whether each of the following is the graph of a function.

2.   3.   4.

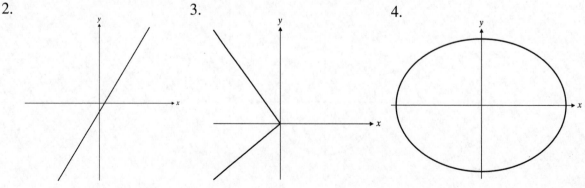

Introduction to Functions
Topic 9.3 Function Notation

Vocabulary
function notation   •   domain   •   range

1. If the name of a function is f and the variable is x, the function can be represented by

    the _____ $f(x)$.

Step-by-Step Video Notes
Watch the Step-by-Step Video lesson and complete the examples below.

Example	Notes
1 & 2. Use function notation to rewrite the following functions using the given function names. $y = 9x - 2$, function name f $y = -16t^2 + 10$, function name h	
3. Determine the domain and range of the function. $$f(x) = -4x + 10$$ Answer:	

Example	Notes

4 & 5. Determine the domain and range of each function.

$$g(x) = x^2 - 4$$

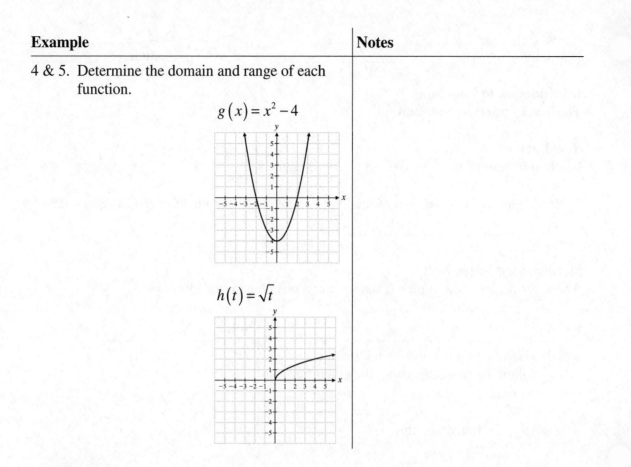

$$h(t) = \sqrt{t}$$

Helpful Hints
Function notation is useful because the variable that makes up the domain is easily identified as the variable inside the parentheses. Read $f(x)$ as "f of x." It does not mean f times x.

Concept Check
1. A function g is defined as $x + y = 3$. Rewrite using the function notation $g(x)$.

Practice
Use function notation to rewrite the functions using the given function names.

2. $y = 3x - 7$, function name f

3. $y = -3x^2 + 5$, function name g

Determine the domain and range of the function based on its graph

4. $f(x) = \dfrac{4}{3}x - 3$

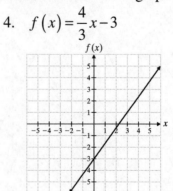

Introduction to Functions
Topic 9.4 Evaluating Functions

Vocabulary
function notation   •   evaluating a function   •   coordinate

1. When _____ at a certain value, substitute that value for the variable in the expression and simplify.

Step-by-Step Video Notes
Watch the Step-by-Step Video lesson and complete the examples below.

Example	Notes
1 & 2. If $f(x) = x + 8$, find the following. $f(2)$ $f(-6)$	
3–5. If $f(x) = 2x^2 - 4$, find each of the following. $f(5)$ $f(-3)$ $f(0)$	

Example	Notes
6. The approximate length of a man's femur (thigh bone) is given by the function $f(x) = 0.5x - 17$, where x is the height of the man in inches. Find the approximate length of the femur of a man who is 70 inches tall. Substitute 70 for x. Simplify. Answer:	

Helpful Hints

To find a function's value when x is some number a, we write $f(a)$. This point is shown on a graph by the coordinate $(a, f(a))$.

When evaluating a function, it is helpful to place parentheses around the value that is being substituted for x.

When a function defines a real-world relationship, be sure to use the correct units.

Concept Check

1. If $f(x) = (x-2)^2 - 9$, is $f(2) = f(-2)$? Is $f(5) = f(-1)$?

Practice

If $f(x) = 3x - 7$, find the following.

2. $f(-2)$

3. $f(2.6)$

If $h(t) = -16t^2 + 500$, find the following.

4. $h(3)$

5. $h(5)$

Introduction to Functions
Topic 9.5 Piecewise Functions

Vocabulary
independent variable   •   piecewise function   •   absolute value

1. A(n) _____ is a function whose definition changes depending on the value of the independent variable.

Step-by-Step Video Notes
Watch the Step-by-Step Video lesson and complete the examples below.

Example	Notes
1. If $f(x) = \begin{cases} 2x & \text{if} & x < 2 \\ x^2 & \text{if} & x \geq 2 \end{cases}$, find the following. $f(3)$ Determine which domain the value fits into, then substitute 3 for x in the corresponding expression and simplify. Answer:	
2. If $f(x) = \begin{cases} 2x+3 & \text{if} & x \leq 0 \\ -x-1 & \text{if} & x > 0 \end{cases}$, find the following. $f(2)$ $f(-6)$ $f(0)$	

Example	Notes
3. Susan's weekly salary (in dollars) is given by $f(x) = \begin{cases} 10x & \text{if } 0 \leq x \leq 40 \\ 20x - 400 & \text{if } x > 40 \end{cases}$, where x is the number of hours worked per week. Find how much money Susan will make if she works the following number of hours in a week. 30 hours 60 hours	
4. Write the absolute value function $f(x) = |x|$ as a piecewise function.	

Helpful Hints

Each x value of a function must have one and only one y value. Do not substitute the value of x into more than one expression when evaluating a function.

Concept Check

1. Is $f(x) = \begin{cases} x^2 & \text{if } x < 0 \\ 4x^2 - 3x^2 & \text{if } x \geq 0 \end{cases}$ a piecewise function? Explain.

Practice

Find the value of $f(2)$ for each of the following piecewise functions.

2. $f(x) = \begin{cases} 5x - 7 & \text{if } x \leq 2 \\ -x + 4 & \text{if } x > 2 \end{cases}$ 3. $f(x) = \begin{cases} x^2 + 6 & \text{if } x \leq 0 \\ -x^2 - 1 & \text{if } x > 0 \end{cases}$ 4. $f(x) = \begin{cases} 0.5x & \text{if } x > 0 \\ x^{16} & \text{if } x \leq 0 \end{cases}$

Solving Systems of Linear Equations
Topic 10.1 Introduction to Systems of Linear Equations

Vocabulary
system of linear equations   •   ordered pair   •   solution to a system of linear equations

1. A _____ is a set of two or more linear equations containing the same variables.

Step-by-Step Video Notes
Watch the Step-by-Step Video lesson and complete the examples below.

Example	Notes
1. Determine whether $(3,-2)$ is a solution to the following system of equations. $$x+3y=-3$$ $$4x+3y=6$$ Answer:	
2. Determine whether $(4,3)$ is a solution to the following system. $$7x-4y=16$$ $$5x+2y=24$$ Answer:	

Example	Notes
4. Determine whether $\left(\dfrac{4}{3}, \dfrac{1}{6}\right)$ is a solution to the following system. $2y = 1 - \dfrac{1}{2}x$ $3x = 2 + 12y$ Answer:	

Helpful Hints

If the point (x, y) exists on two lines, then it is a solution to the system of the equations which contains those lines.

Concept Check

1. If $(2,1)$ is the solution to a system of two equations and a third equation is added to the system, is $(2,1)$ still a solution? Explain.

Practice

Determine whether $(4, -1)$ is a solution to the following systems.

2. $x + 2y = 2$
   $5x + 3y = 16$

Determine whether $(0.5, 3)$ is a solution to the following systems.

4. $2x = 3y - 1$
   $\dfrac{1}{3}y = 6x - 2$

3. $3x - 4y = 16$
   $5x + 6y = 14$

5. $0.6x + 0.4y = 1.5$
   $8x + 3y = 12$

Solving Systems of Linear Equations
Topic 10.2 Solving by the Graphing Method

Vocabulary
inconsistent system of equations   •   dependent system of equations   •   point of intersection

1. A(n) _____ has no solution. Its graph will be two parallel lines, which do not intersect. There are no ordered pairs in common.

2. A(n) _____ has an infinite number of solutions. This means that the graphs of the two equations will show the same line.

Step-by-Step Video Notes
Watch the Step-by-Step Video lesson and complete the examples below.

Example	Notes
1. Find the solution (the point of intersection of the two lines).	

$3x + y = 2$

$2x - y = 3$

Solution:

Example	**Notes**

3. Find the solution to the system of equations by graphing.

$$y = x - 1$$
$$2x + y = 8$$

Answer:

Helpful Hints

When you solve a system of linear equations by graphing, graph the equations on the same graph. Find the solution, which is the point of intersection of the lines. Write the solution as an ordered pair. Verify your solution by substituting the ordered pair into both equations.

Concept Check

1. What are the three possible types of solutions to a system of linear equations?

Practice

Find the solution to the system of equations by graphing.

2. $y = x + 1$
   $y = 3x - 1$

3. $x + y = 5$
   $y = \dfrac{1}{2}x - 1$

4. $2x + 3y = 9$
   $y = -2x + 3$

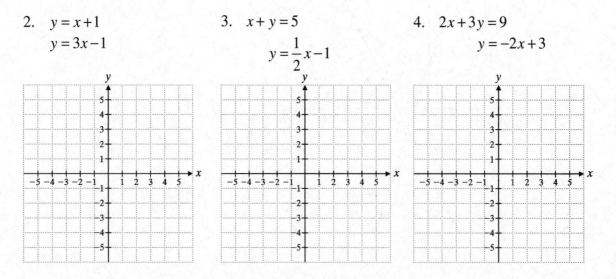

Solving Systems of Linear Equations
Topic 10.3 Solving by the Substitution Method

Vocabulary
substitution method   •   system of equations   •   no solution

1. The _____ involves choosing either equation in a system and solving for either variable and substituting the result into the other equation to solve for the remaining variable.

Step-by-Step Video Notes
Watch the Step-by-Step Video lesson and complete the examples below.

Example	Notes
1. Solve the system of equations by substitution. $$4x + 3y = 50$$ $$y = 2x$$ Substitute $2x$ for y in the first equation. Simplify and solve for x. Use this value for x in one equation to find y. Check your solution in the other equation. Solution:	
2. Solve the system of equations by substitution. $$y = x - 5$$ $$3x - 2y = 17$$ Solution:	

Example	Notes
4. Solve the system of equations by substitution. Solve one equation for one variable. $4x - 24y = 40$ $x - 6y = 10$ Solution:	
5. Solve the system of equations by substitution. $x + 3y = 5$ $4x + 12y = 40$ Solution:	

Helpful Hints

If one of the equations has a variable with a coefficient of 1 or -1, choose that equation to solve. This will almost always make this step easier.

Concept Check

1. Describe a situation where solving a system by substitution would be easier than solving it by graphing.

Practice

Solve the system of equations by substitution.

2. $4x + 3y = 38$
   $\quad\quad y = 5x$

4. $3x + 4y = 8$
   $\quad 2x + y = -3$

3. $6x + 5y = 10$
   $\quad\quad x = y + 9$

5. $\quad\quad y = 4x + 7$
   $\quad 8x - 2y = 19$

Solving Systems of Linear Equations
Topic 10.4 Solving by the Elimination Method

Vocabulary

Elimination Method   •   many solutions   •   no solution

1. When solving systems of equations, you can use the _____ to solve the system if the coefficients of either variable in the equations are opposites.

Step-by-Step Video Notes

Watch the Step-by-Step Video lesson and complete the examples below.

Example	**Notes**
1. Solve the system of equations. $7x - 3y = 1$ $5x + 3y = 11$ Solution:	
2. Solve the system of equations by the Elimination Method. $3x + 7y = 22$ $2x - 7y = 3$ Check to see if the coefficients of either variable are opposites. Solution:	

Example	Notes
4. Solve the system of equations by the Elimination Method. $4x + 3y = -5$ $7x + 2y = 14$ Solution:	
5. Solve the system of equations by the Elimination Method. $3x + 2y = 18$ $-3x - 2y = 14$ Solution:	

Helpful Hints

An easy way to use elimination is to pick which variable you want to eliminate and multiply each equation by the coefficient of that variable term in the other equation. It may be necessary to also multiply by a negative number to produce opposite variable terms.

Concept Check

1. Describe a situation where solving a system by elimination would be easier than solving by substitution.

Practice

Solve the system by the Elimination Method.

2. $7x + 4y = 8$
   $-7x - 6y = 2$

4. $7x + 5y = 98$
   $8x - 2y = 58$

3. $5x - 6y = 17$
   $2x - 4y = 6$

5. $48x - 33y = 57$
   $-16x + 11y = 19$

Name: _____   Date: _____

Instructor: _____   Section: _____

Solving Systems of Linear Equations
Topic 10.5 Applications of Systems of Linear Equations

Vocabulary
system of equations   •   matrix   •   word problem

1. To solve a _____ using a _____, read the question and determine what information you will need to solve the problem.

Step-by-Step Video Notes
Watch the Step-by-Step Video lesson and complete the examples below.

Example	Notes
1. A movie theater sells tickets for $10 and bags of popcorn for $3. In a single Saturday night, the theater had $2375 in sales. The theater owner found that if he raised ticket prices to $11 and the popcorn prices to $4 and sold the same number of tickets and popcorn bags, the theater would make $2700. How many tickets were sold? How many bags of popcorn were sold? Answer:	
2. A boat travels 20 miles upstream, against the current, in 4 hours. The return trip, 20 miles downstream, with the current, takes only 2 hours. Find the speed of the boat in still water and the speed of the current. Answer:	

Example	Notes
3. An electronics company makes two types of switches. Type A takes 4 minutes to make and requires $3 worth of materials. Type B takes 5 minutes to make and requires $5 of materials. In the latest production batch, it took 35 hours to make these switches, and the materials cost $1900. How many of each type of switch were made?	

Answer:

Helpful Hints

Remember that once you solve a system relating to an application, it is important to go back and answer the question asked. This requires interpreting the solution to the system.

When you produce a system relating to an application, you can solve it using any method. Choose the method that is easiest for you or that is easiest to use with the particular system that you found.

Concept Check

1. While calculating the answer to Example 2, Tia incorrectly finds that the speed of the current is 0 mph and the speed of the boat is 20 mph. Explain her error.

Practice

2. A plane travels 3150 miles with the wind, in 7 hours. The return trip, 3150 miles against the wind, takes 9 hours. What is the speed of the plane?

3. What is the speed of the wind in Practice 2?

4. Jim charges $40 to mow lawns and it takes him 3 hours to mow each one. He also charges $15 to weed gardens, which takes him an hour. In a month Jim worked 160 hours and made $2175. How many lawns did Jim mow?

5. How many gardens did Jim weed in Practice 4?

Introduction to Polynomials and Exponent Rules
Topic 11.1 Introduction to Polynomials

Vocabulary
term • coefficient • like terms • polynomial • descending order • monomial
degree of a term • degree of a polynomial • binomial • trinomial • exponent

1. The _____ is the highest exponent of the base in that term. If there is more than one variable, it is the sum of the exponents of all of the variables in the term.

2. A _____ in the variable x is the sum of a finite number of terms of the form ax^n, where a is any real number and n is a whole number.

Step-by-Step Video Notes
Watch the Step-by-Step Video lesson and complete the examples below.

Example	**Notes**
2. Find the degree of the term $4ab^2$. The degree of a term is the highest exponent of the base in that term. This term has two variables, ____ and ____. The exponent of ____ is ⬚, and of ____ is ⬚. Answer:	
4. For the polynomial $5x^3 + 8x^2 - 20x - 2$, find the degree of each term. Then find the degree of the polynomial. The degree of the first term is ⬚, the degree of the second term is ⬚, the degree of the third term is ⬚, and the degree of the last term is ⬚ since it is a constant. Answer:	

Example	Notes
6. State the degree of the polynomial $5x + 3x^3$, and whether it is a monomial, a binomial, or a trinomial. Answer:	
9. Evaluate the polynomial $P(x) = 2x^2 - 6x + 7$ at $x = -2$. Answer:	

Helpful Hints

You can combine like terms by adding or subtracting the coefficients of the terms.

A polynomial in x is said to be written in descending order if it is written so that the exponents on the variable x decrease from left to right.

The degree of a polynomial is the highest degree of any of its terms. The polynomial 0 is said to have no degree. A polynomial consisting of a constant only is said to have degree 0.

Concept Check

1. Is the polynomial $24x^3 + 8x^2 + 4x + 2$ of greater degree than the monomial $-2x^5$?

Practice

State the degree of the polynomial, and tell if it is a monomial, binomial, or trinomial.

2. $72x^3 + 45x^2 + 27x$

3. $15x^4y^2 - 22x^3y$

Evaluate the polynomial at the given value of the variable.

4. $f(x) = 2x^2 + 3x - 9$ for $x = 1.5$

5. $h(t) = -16t^2 + 400$ for $t = 4$

Introduction to Polynomials and Exponent Rules
Topic 11.2 Addition of Polynomials

Vocabulary

adding polynomials   •   coefficients   •   like terms

1. When _____, combine like terms.

Step-by-Step Video Notes
Watch the Step-by-Step Video lesson and complete the examples below.

Example	Notes
2. Add. $$\left(5x^2 - 6x - 12\right) + \left(-3x^2 - 9x + 5\right)$$ Remove parentheses and identify like terms. $$5x^2 - 6x - 12 - 3x^2 - 9x + 5$$ Combine like terms. Answer:	
3. Add. $$\left(7x^2 + 8x + 9\right) + \left(13x^2 - 10x + 5\right)$$ Remove parentheses and identify like terms. Combine like terms. We can also use a vertical format to help visualize the addition. $$7x^2 + \ 8x + 9$$ $$+13x^2 - 10x + 5$$ Answer:	

Example	Notes
4. Add. $\left(1.2x^3 - 5.6x^2 + 5\right) + \left(-3.4x^3 - 1.2x^2 + 4.5x - 7\right)$ Answer:	
5. Add. $\left(\dfrac{1}{2}x^2 - 6x + \dfrac{1}{3}\right) + \left(2x - \dfrac{1}{2} + \dfrac{1}{5}x^3\right)$ Answer:	

Helpful Hints

You can add polynomials by combining like terms. While each polynomial may not be given in descending order, arrange the sum in descending order.

If you use a vertical format to add polynomials, be sure each column contains terms of the same degree.

Concept Check

1. How many terms are in the polynomial sum $\left(4x^3 - 7x^2 + 3x\right) + \left(-2x^2 + 9x + 3\right)$?

Practice

Add.

2. $\left(-4x^3 + 3x^2 - 2\right) + \left(7x^3 - 8x^2 - 3\right)$

4. $\left(-\dfrac{1}{2}x^2 - 5x - 6\right) + \left(8x + \dfrac{3}{4}x^2 + \dfrac{2}{3}\right)$

3. $\left(4.4x^3 - 0.13x^2 + 2.11x\right) + \left(-0.07x^2 - 1.89x + 6\right)$

5. $\left(13x^2 + 2 - 5x\right) + \left(13x - 9x^2 + 7\right)$

Name: _____   Date: _____

Instructor: _____   Section: _____

Introduction to Polynomials and Exponent Rules
Topic 11.3 Subtraction of Polynomials

Vocabulary
Adding polynomials   •   subtracting polynomials   •   like terms

1. When _____, change the sign of each term in the second polynomial and add the result to the first polynomial by combining like terms.

Step-by-Step Video Notes
Watch the Step-by-Step Video lesson and complete the examples below.

Example	Notes
1. Subtract. $$(2x+3)-(x-5)$$ Change the sign of each term in the second polynomial and add. $$2x+3-\boxed{}+\boxed{}$$ Add the polynomials by combining like terms. $$2x-\boxed{}+3+\boxed{}$$ Answer:	
2. Subtract. $$\left(-2x^3+7x^2-3x-1\right)-\left(-6x^3-9x^2-x+4\right)$$ Change the sign of each term in the second polynomial and add. You can also use a vertical format to help visualize the addition. $$\begin{array}{l}-2x^3+7x^2-3x-1\\ \underline{-\left(-6x^3-9x^2\ -x+4\right)}\end{array}\ \rightarrow\ \begin{array}{l}-2x^3\ +7x^2\ -3x\ -1\\ +\boxed{}\end{array}$$ Answer:	

Example	Notes
3. Subtract. $\left(-3x^4 + 5x^2 + 2\right) - \left(6x^3 - 10x^2 + 2x - 1\right)$ Answer:	
5. Subtract. $\left(-6x^2y - 3xy + 7xy^2\right) - \left(5x^2y - 8xy - 15x^2y^2\right)$ Answer:	

Helpful Hints

Use extra care in determining which terms are like terms when polynomials contain more than one variable. Every exponent of every variable in the two terms must be the same if the terms are to be like terms.

If you use a vertical format to subtract polynomials, be sure to change the sign of each term in the second polynomial before adding.

Concept Check

1. How many terms are in the difference $\left(4xy - 7x^2y^2 + 3y\right) - \left(-2x^2y^2 + 9xy + 3\right)$?

Practice
Subtract.

2. $\left(4x^2 + 6x - 9\right) - \left(-5x^2 - 6x + 2\right)$

4. $\left(6.3x^2 - 1.8x + 3.5\right) - \left(3.2x - 0.7x^2 - 4.9\right)$

3. $\left(15x^5 - 8x^3 + 5x - 3\right) - \left(-5x^5 + 12x^4 + 6x^2\right)$

5. $\left(a^4 - 7ab + 3ab^2 - 2b^3\right) - \left(2a^4 + 4ab - 6b^3\right)$

Introduction to Polynomials and Exponent Rules
Topic 11.4 Product Rule for Exponents

Vocabulary

exponential expression   •   product rule for exponents   •   base

1. A variable or a number raised to an exponent is a(n) _____.

2. The _____ states that to multiply two exponential expressions that have like bases, keep the base and add the exponents, or $x^a \cdot x^b = x^{a+b}$.

Step-by-Step Video Notes

Watch the Step-by-Step Video lesson and complete the examples below.

Example	Notes
3. Multiply $x^3 \cdot x \cdot x^6$. The base of each factor is $\boxed{}$. Keep this base and add the exponents. Note that even though it is not written, the exponent of the term x is $\boxed{}$. $x^3 \cdot x \cdot x^6 = x^{\left(\boxed{} + \boxed{} + \boxed{}\right)}$ Answer:	
5. Simplify $2^3 \cdot 2^5$, if possible. Write your answer using exponential notation. Answer:	

Example	Notes
7. Multiply $(3a)^2 \cdot (3a)^4$. The base of each factor is ☐ . Answer:	
10. Multiply $(5ab)\left(-\dfrac{1}{3}a\right)(9b^2)$. Answer :	

Helpful Hints

It is important that you apply the product rule even when the exponent is 1. Every variable that does not have a written exponent is understood to have an exponent of 1.

To multiply exponential expressions with coefficients, first multiply the coefficients, and then multiply the variables with exponents separately.

Concept Check

1. Can you use the product rule for exponents to simplify $p^2 \cdot g^2$? Explain.

Practice
Multiply.

2. $w^{12} \cdot w$

3. $\left(-8x^4\right)\left(-5x^3\right)$

4. $(4xy)\left(-\dfrac{1}{8}x^4y^2\right)\left(4xy^4\right)$

5. $\left(-6.5pq^5\right)\left(-p^3\right)\left(2p^4q\right)$

Name: _____   Date: _____

Instructor: _____   Section: _____

Introduction to Polynomials and Exponent Rules
Topic 11.5 Power Rule for Exponents

Vocabulary
product rule for exponents   •   quotient raised to a power   •   exponential expression
power rule for exponents   •   product raised to a power   •   exponential notation

1. The _____ states that $\left(x^a\right)^b = x^{ab}$.

2. The rule for a _____ is demonstrated by the equation $\left(\dfrac{x}{y}\right)^a = \dfrac{x^a}{y^a}$,

    if $y \neq 0$.

Step-by-Step Video Notes
Watch the Step-by-Step Video lesson and complete the examples below.

Example	**Notes**
1–3. Simplify. Write your answer using exponential notation. $\left(x^3\right)^5$ $\left(2^7\right)^3$ $\left(y^2\right)^4$	
4–6. Simplify. $\left(ab\right)^8$ $\left(3x\right)^4$ $\left(-2x^2\right)^3$	

Example	Notes
8. Simplify $\left(\dfrac{3}{w}\right)^4$. Write your answer using exponential notation. Answer:	
9. Simplify $\left(\dfrac{-3x^2z^0}{y^3}\right)^4$. Answer:	

Helpful Hints

If a product in parentheses is raised to a power, the parentheses indicate that each factor inside the parentheses must be raised to that power.

When simplifying expressions by using multiple rules involving exponents, be careful determining the correct sign, especially if there is a negative coefficient.

Concept Check

1. A student simplified $\left(-3x^2\right)^4$ as $-81x^8$. Is this correct? If not, what is the error?

Practice
Simplify.

2. $\left(d^4\right)^5$

3. $\left(-8x^4\right)^2$

4. $\left(\dfrac{p}{2}\right)^5$

5. $\left(-\dfrac{2y^0z^3}{3x^5}\right)^4$

Multiplying Polynomials
Topic 12.1 Multiplying by a Monomial

Vocabulary
Distributive Property   •   product rule for exponents   •   monomial

1. A _____ is a polynomial with exactly one term.

2. To multiply a monomial by a polynomial, use the _____ to multiply the
    monomial by each term in the polynomial.

Step-by-Step Video Notes
Watch the Step-by-Step Video lesson and complete the examples below.

Example	Notes
1. Multiply $3x^2(5x-2)$. Answer:	
2. Multiply $2x(x^2+3x-1)$. Multiply the monomial by each term in the parentheses. Answer:	

Example	Notes
3. Multiply $-6xy\left(x^3 + 2x^2y - y^2\right)$. Answer:	
5. Multiply $\left(2x^2 - 3x + 8\right)\left(-7x\right)$. Use the Commutative Property to write the monomial first. Answer :	

Helpful Hints

Remember with a term such as $7x$ that the exponent on x is 1. When you multiply, add the powers, so $7x \cdot x^2 = 7x^{(1+2)} = 7x^3$.

Multiplication is commutative. The order in which terms are multiplied doesn't matter. A monomial can be after a polynomial, but a monomial is usually written before a polynomial.

Concept Check

1. Which multiplication would you perform first to simplify $\left(2x\right)\left(3y\right)\left(x^2 + 4y\right)$? Why?

Practice

Multiply.

2. $6x^3\left(-3x^2 + 2x\right)$

3. $5x\left(x^2 - 4x - 7\right)$

4. $-3xy^2\left(x^2 + 8xy - 4y^2\right)$

5. $\left(x^3 - 6x + 12\right)\left(-3x\right)$

Multiplying Polynomials
Topic 12.2 Multiplying Binomials

Vocabulary
binomial   •   FOIL method   •   polynomial

1. The _____ for multiplying two binomials means multiply the first terms, the outer terms, the inner terms, and then the last terms.

2. A _____ is a polynomial with exactly two terms.

Step-by-Step Video Notes
Watch the Step-by-Step Video lesson and complete the examples below.

Example	Notes
1. Multiply $(3x+1)(x+4)$. Multiply each term in the first binomial by each term in the second binomial. $3x\left(\boxed{}\right)+3x\left(\boxed{}\right)+1\left(\boxed{}\right)+1\left(\boxed{}\right)$ Combine like terms to simplify the expression. Answer:	
3. Multiply $(4x-9y)(8x-3)$. Answer:	

Example	Notes
4. Multiply $(x+3)(x+5)$ using the FOIL method.	
Multiply the first terms.	
Multiply the outer terms.	
Multiply the inner terms.	
Multiply the last terms.	
Answer:	
6. Multiply $(-2+7x)(-9x+5)$ using the FOIL method.	
Answer:	

Helpful Hints

The FOIL method is just a way to help you remember how to multiply binomials. It is only used to multiply a binomial by a binomial.

Concept Check

1. Would you use the FOIL method when multiplying $(7x^2-3x^2)(-4x^3+18y)$? Why or why not?

Practice

Multiply.

2. $(x+4)(3x+7)$

3. $(5x-12y)(9x-4)$

Multiply using the FOIL method.

4. $(x^2+3)(x^2+8)$

5. $(4x-6)(-3x-2)$

Multiplying Polynomials
Topic 12.3 Multiplying Polynomials

Vocabulary
polynomial   •   multiplying polynomials   •   Distributive Property

1. When _____, multiply each term in the first polynomial by every term in the second polynomial. Then write the sum of the products in descending order.

Step-by-Step Video Notes
Watch the Step-by-Step Video lesson and complete the examples below.

Example	Notes
1. Multiply $(x+4)(3x^2+x+2)$. Multiply each term in the first polynomial by every term in the second polynomial. Combine like terms. Write the final polynomial in descending order. Answer:	
2. Multiply $(3x-1)(6x^2-5x+8)$. Answer:	

Example	Notes
4. Multiply $\left(4x^2+9x+7\right)\left(2x^2-6x-5\right)$.	
Answer:	
5. Multiply $\left(x+1\right)\left(x+2\right)\left(x+3\right)$. Multiply the first two polynomials, then multiply the product by the third polynomial.	

Answer:

Helpful Hints

A good way to keep terms organized when multiplying bigger polynomials is to multiply the polynomials vertically. This is like multiplying multi-digit numbers vertically.

You can also multiply three or more polynomials; just multiply them two at a time.

Concept Check

1. Which multiplication would you perform first to simplify $\left(x+4\right)\left(x^2+7-3x\right)\left(x+1\right)$? Why?

Practice
Multiply.

2. $\left(8x+4\right)\left(x^2+7x-6\right)$

3. $\left(5x^2-9x+4\right)\left(6x-2\right)$

4. $\left(x^2+6x+8\right)\left(2x^2-x-6\right)$

5. $\left(4x-2\right)\left(3x+7\right)\left(x+5\right)$

Multiplying Polynomials
Topic 12.4 Multiplying the Sum and Difference of Two Terms

Vocabulary
FOIL method   •   multiplying binomials: a sum and a difference

1. The rule _____ states that $(a+b)(a-b)=a^2-b^2$, where a and
   b are numbers or algebraic expressions.

Step-by-Step Video Notes
Watch the Step-by-Step Video lesson and complete the examples below.

Example	Notes
1. Multiply $(5x+4)(5x-4)$. Multiply each term in the first polynomial by each term in the second polynomial. Answer:	
2. Multiply $(x+5)(x-5)$. Answer:	

Example	Notes
4 & 5. Multiply.	

$$\left(2x^2 + 3y\right)\left(2x^2 - 3y\right)$$

$$\left(\frac{1}{4}x - \frac{2}{3}\right)\left(\frac{1}{4}x + \frac{2}{3}\right)$$

Helpful Hints

Remember that the sum and the difference have to be of the same terms. For example, $(a+b)(a-b) = a^2 - b^2$, but $(a+b)(c-d) \neq ac - bd$.

Remember, when squaring a fraction, you square the numerator and square the denominator.

Concept Check

1. Multiply $(x+y)(x-y)(x^2+y^2)$ without using the FOIL method.

Practice

Multiply.

2. $(x+4)(x-4)$

4. $\left(8x^2 + 11y\right)\left(8x^2 - 11y\right)$

3. $(6x+9)(6x-9)$

5. $\left(\frac{4}{5}n - \frac{2}{7}\right)\left(\frac{4}{5}n + \frac{2}{7}\right)$

Multiplying Polynomials
Topic 12.5 Squaring Binomials

Vocabulary
a binomial squared   •   binomial   •   FOIL method

1. The rule _____ states that $(a+b)^2 = a^2 + 2ab + b^2$ and

   $(a-b)^2 = a^2 - 2ab + b^2$ where a and b are numbers or algebraic expressions.

Step-by-Step Video Notes
Watch the Step-by-Step Video lesson and complete the examples below.

Example	Notes
1. Simplify $(2x-3)^2$. Write the expression as a multiplication problem, then multiply the binomials. Answer:	
2. Simplify $(3x+5)^2$. Answer:	

119

Example	Notes
4. Simplify $\left(2x^2 + 3y\right)^2$.	
Answer:	

5. Simplify $\left(\dfrac{1}{4}x - \dfrac{2}{3}\right)^2$.

Answer :

Helpful Hints

When you square a binomial, the middle term of the product will always be double the product of the terms of the binomial. The product is called a perfect square trinomial.

The Binomial Squared rule is helpful because it is only necessary to find the squares of the two terms in the binomial and twice their product. Using the FOIL method yields the same result, but requires more calculation and simplification.

Concept Check

1. What will be the middle term when you square the binomial $\left(3x - 4\right)$?

Practice
Simplify.

2. $\left(x + 7\right)^2$

3. $\left(5x - 9\right)^2$

4. $\left(14x^3 + 13y\right)^2$

5. $\left(\dfrac{3}{7}y - \dfrac{1}{11}\right)^2$

Dividing Polynomials and More Exponent Rules
Topic 13.1 The Quotient Rule

Vocabulary
quotient rule   •   prime factors method   •   Zero as an Exponent Property

1. The _____ states that for all non-zero numbers x, $\dfrac{x^a}{x^b} = x^{a-b}$.

2. The _____ states that for all non-zero numbers x, $\dfrac{x^a}{x^a} = x^0 = 1$.

Step-by-Step Video Notes
Watch the Step-by-Step Video lesson and complete the examples below.

Example	Notes
2. Simplify $\dfrac{25x^6}{10x^3}$. Write the numerator and denominator in an expanded form, to show the factors. Answer:	
5. Simplify $-\dfrac{16s^6}{32s^2}$. Divide the number parts by the GCF, then use the quotient rule to simplify the variable part. Answer:	

121

Example	Notes
6. Simplify $\dfrac{7^{10}}{7^4}$ using the quotient rule. Answer:	
9. Evaluate $\dfrac{(ax)^4}{(ax)^4}$. Answer:	

Helpful Hints

The quotient rule says to divide like bases, keep the base and subtract the exponents.

Any non-zero number raised to an exponent of zero is equal to 1. 0^0 is undefined.

Concept Check

1. Simplify $\dfrac{a^3}{a^0}$ in two different ways.

Practice

Simplify by the prime factors method.

2. $\dfrac{x^8}{x^4}$

3. $-\dfrac{42x^8}{77x^5}$

Simplify using the quotient rule.

4. $\dfrac{x^{49}}{x^7}$

5. $\dfrac{(37np)^9}{(37np)^9}$

Dividing Polynomials and More Exponent Rules
Topic 13.2 Integer Exponents

Vocabulary
negative exponent   •   power rule   •   product rule

1. By definition of a _____, $x^{-n} = \dfrac{1}{x^n}$, if $x \neq 0$.

Step-by-Step Video Notes
Watch the Step-by-Step Video lesson and complete the examples below.

Example	Notes
1–3. Simplify. Write your answers with positive exponents. z^{-6} $\dfrac{x^3}{x^7}$ $\left(x^{-5}\right)\left(x^3\right)$	
4–6. Simplify. Write your answers with positive exponents. 2^{-5} -5^{-2} $(-3)^{-3}$	

Example	Notes
8. Simplify $\dfrac{x^{-4}}{y^{-2}}$. Write your answer with positive exponents. Answer:	
11. Simplify $\dfrac{x^2 y^{-4}}{x^{-5} y^3}$. Write your answer with positive exponents. Answer :	

Helpful Hints

A negative exponent moves the base from one part of the fraction to the other. That is, it moves a numerator to the denominator, and a denominator to the numerator, and the exponent becomes positive.

Concept Check

1. Which is greater, $\left(\dfrac{1}{2}\right)^1$, or $\left(\dfrac{1}{2}\right)^{-1}$?

Practice

Simplify. Write your answers with positive exponents.

2. $2x^{-7}$

3. $\dfrac{x^{-3}}{x^{-5}}$

Evaluate. Write your answers with positive exponents.

4. $\dfrac{(2x)^{-5}}{y^{-10}}$

5. $\dfrac{-3m^{-3}\left(5n^4\right)^{-2}}{m^{-12}n^7}$

Dividing Polynomials and More Exponent Rules
Topic 13.3 Scientific Notation

Vocabulary
Powers of ten   •   scientific notation   •   decimal notation

1. A positive number is written in _____ when it is in the form $a \times 10^n$, where $1 \le a < 10$ and n is an integer.

Step-by-Step Video Notes
Watch the Step-by-Step Video lesson and complete the examples below.

Example	Notes
1. Write $23{,}400{,}000$ in scientific notation. Move the decimal point to put one non-zero digit to the left of the decimal point. 2 3 4 0 0 0 0 0 The point was moved ☐ places. Answer:	
3. Write 2.31×10^6 in decimal notation. The positive power of 10 means move the decimal point ☐ places to the _____. Answer:	
6. Write the number in the following application in decimal notation. The volume of a gold atom is 1.695×10^{-23} cubic centimeters. Write this in decimal notation. Answer:	

Example	Notes
7 & 8. Simplify.	

$$\left(8\times10^{16}\right)\left(7\times10^{4}\right)$$

$$\frac{1.5\times10^{7}}{2.5\times10^{2}}$$

Helpful Hints
Write numbers in scientific notation to make a very big or very small number more compact.

To write a number in decimal form, move the decimal point the correct number of places in the appropriate direction. If the power of 10 is positive, the decimal point moves right. If the power of 10 is negative, the decimal point moves left.

Concept Check
1. How many zeros are at the end of the decimal number equivalent to 3.048×10^{7}?

Practice
Write each number in scientific notation.

2. $24,300,000$

Write in decimal notation.

4. 3.04×10^{9}

3. $\left(2.5\times10^{-4}\right)\left(4\times10^{6}\right)$

5. 2.763×10^{-8}

Dividing Polynomials and More Exponent Rules
Topic 13.4 Dividing a Polynomial by a Monomial

Vocabulary
trinomial   •   monomial   •   polynomial

1. When dividing a polynomial by a monomial, divide each term of the _____ by the
   _____.

Step-by-Step Video Notes
Watch the Step-by-Step Video lesson and complete the examples below.

Example	Notes
3. Divide $\dfrac{15x^5 + 10x^4 + 25x^3}{5x^2}$. Divide each term in the polynomial by $5x^2$. Write the sum of the results. Answer:	
4. Divide $\dfrac{63x^7 - 35x^6 - 49x^5}{-7x^3}$. Answer:	

Example	Notes
5. Divide $\left(36x^3 - 18x^2 + 9x\right) \div \left(9x\right)$. Answer:	
7. Divide $\dfrac{24x^3 + 16x^2 - 56x}{8x^2}$. Answer :	

Helpful Hints

Split up a fraction into an addition of two or more fractions when trying to simplify the fraction, or with division of a polynomial by a monomial.

If each term does not divide evenly, simplify each individual fraction.

Concept Check

1. Biff states incorrectly that $\left(24x^3 + 8x^2\right) \div \left(8x^2\right) = 3x$. What was his error?

Practice
Divide.

2. $\dfrac{72x^8 + 45x^6 + 27x^4}{9x^2}$

4. $\dfrac{-42x^9 + 24x^7 + 18x^5}{-6x^5}$

3. $\left(150x^4 - 220x^3 + 90x^2\right) \div \left(10x^2\right)$

5. $\dfrac{-33x^8 - 24x^5 + 9x^4}{-9x^4}$

Dividing Polynomials and More Exponent Rules
Topic 13.5 Dividing a Polynomial by a Binomial

Vocabulary
long division   •   polynomial long division   •   quotient

1. When setting up _____, place the terms of the polynomials in descending order. Insert a zero for any missing terms.

Step-by-Step Video Notes
Watch the Step-by-Step Video lesson and complete the examples below.

Example	Notes
1. Divide $\left(6x^2 + 7x + 2\right) \div \left(2x + 1\right)$. Set the problem up using the long-division symbol. Divide $6x^2$ by $2x$. This is the first term of the answer. $$\begin{array}{r} \boxed{3x} \\ 2x+1{\overline{\smash{\big)}\,6x^2+7x+2}} \\ 6x^2+3x \end{array}$$ Answer:	
2. Divide $\left(x^3 + 5x^2 + 11x + 4\right) \div \left(x + 2\right)$. $$\begin{array}{r} \boxed{} \\ x+2{\overline{\smash{\big)}\,x^3+5x^2+11x+4}} \end{array}$$ Answer:	

Example	Notes
3. Divide $\left(5x^3 - 24x^2 + 9\right) \div \left(5x + 1\right)$. Answer:	

Helpful Hints

After setting up a polynomial long division problem, divide the first term of the dividend by the first term of the divisor. The result is the first term of the answer. Then proceed as you would with numbers until the degree of the remainder is less than the degree of the divisor.

Concept Check

1. What missing term must you insert in the dividend when dividing $\left(p^3 - p + 8\right) \div \left(p - 4\right)$?

Practice
Divide.

2. $\left(6x^2 + 4x - 10\right) \div \left(3x + 5\right)$ 3. $\left(28x^2 - 15x - 20\right) \div \left(4x + 3\right)$ 4. $\left(n^3 - n^2 - 4\right) \div \left(n - 2\right)$

Factoring Polynomials
Topic 14.1 Greatest Common Factor

Vocabulary
factor   •   greatest common factor   •   factoring a polynomial
prime polynomial   •   distributive property

1. _____ is the process of writing a polynomial as a product or two or more factors.

2. When two or more numbers, variables, or algebraic expressions are multiplied, each is called a _____.

Step-by-Step Video Notes
Watch the Step-by-Step Video lesson and complete the examples below.

Example	**Notes**
1–3. Find the GCF. $x^3, x^7,$ and x^5 $y, y^4,$ and y^7 x and y^2	
5. Factor out the GCF of $9x^5 + 18x^2 + 3x$. Write each term as the product of the GCF and each term's remaining factors. Answer:	

Example	Notes
6. Factor out the GCF of $8x^3y + 16x^2y^2 - 24x^3y^3$.	
Answer:	

7 & 8. Factor out the GCF.

$$24ab + 12a^2 + 36a^3$$

$$3x + 7y + 12xy$$

Helpful Hints
Factoring a polynomial changes a sum and/or difference of terms into a product of factors.

To find the GCF of two or more terms with coefficients and variables, find the product of the GCF of the coefficients and the GCF of the variable factors.

Concept Check
1. Allison states incorrectly that $3x + 24y - 15z$ is a prime polynomial because the variables in the terms are different. What is her error?

Practice
Factor out the GCF.

2. $72x^8 - 54x^6 + 27x^5$

4. $-42x^9y + 24x^7y^2 + 18x^5y^3$

3. $140x^4 - 210x^3 + 70x^2$

5. $-33x^8y^8 - 24x^5y^5 + 9x^4y^3$

Factoring Polynomials
Topic 14.2 Factoring by Grouping

Vocabulary
common factor   •   factoring by grouping   •   polynomial

1. When _____, collect the terms into two groups so that each group
   has a common factor. Then factor out the GCF from each group so that the remaining
   factor in each group is the same.

Step-by-Step Video Notes
Watch the Step-by-Step Video lesson and complete the examples below.

Example	Notes
1. Factor $2x^2 + 3x + 6x + 9$ by grouping. Group terms. Factor within groups. Factor the entire polynomial. Multiply to check your answer. Answer:	
2. Factor $2x^2 + 5x - 4x - 10$ by grouping. Answer:	

Example	Notes
3. Factor $2ax - a - 2bx + b$ by grouping.	
Answer:	
4. Factor $10x^2 - 8xy + 15x - 12y$ by grouping.	
Answer:	

Helpful Hints

Sometimes you will need to factor out a negative common factor from the second two terms to obtain two terms that contain the same binomial factor. When factoring out a negative, check your signs carefully.

Concept Check

1. Can you factor Example 1, $2x^2 + 3x + 6x + 9,$ by grouping the terms differently? If so, do you get the same factors?

Practice
Factor by grouping.

2. $6x^2 - 8x + 9x - 12$

4. $4x + 20y - 3ax - 15ay$

3. $6x + 18y + ax + 3ay$

5. $16a^2 - 14ab + 24a - 21b$

Factoring Polynomials

Topic 14.3 Factoring Trinomials of the Form $x^2 + bx + c$

Vocabulary

trinomial   •   FOIL method   •   prime polynomial

1. A _____ is a polynomial that cannot be factored.

Step-by-Step Video Notes

Watch the Step-by-Step Video lesson and complete the examples below.

Example	Notes
1. Factor $x^2 + 7x + 12$. Write the first two terms of the binomial factors. List the possible pairs of factors of 12. Answer:	
2. Factor $x^2 - 8x + 15$. Write the first two terms of the binomial factors. List the possible pairs of factors of 15. Answer:	

Example	Notes
3. Factor $x^2 - 3x - 10$. Answer:	
5. Factor $4x^2 + 40x - 96$. First factor out the GCF. Answer:	

Helpful Hints

Trinomials of the form $x^2 + bx + c$ factor as $\left(x + \square\right)\left(x + \square\right)$. Trinomials of the form

$x^2 - bx + c$ factor as $\left(x - \square\right)\left(x - \square\right)$, and trinomials of the form $x^2 + bx - c$ factor as

$\left(x + \square\right)\left(x - \square\right)$.

Concept Check

1. To factor a trinomial of the form $x^2 + bx + c$ as the product of two binomials, what must be the product and the sum of the constant terms of the binomial factors?

Practice

Factor.

2. $x^2 + 9x + 18$

4. $3x^2 + 39x + 120$

3. $x^2 - 10x + 24$

5. $x^2 + 13x - 48$

Factoring Polynomials

Topic 14.4 Factoring Trinomials of the Form $ax^2 + bx + c$

Vocabulary

FOIL method • reverse FOIL method

1. To factor the trinomial $ax^2 + bx + c$ using the _____, list the different factorizations of ax^2 and c. List the possible factoring combinations until the correct middle term is reached.

Step-by-Step Video Notes

Watch the Step-by-Step Video lesson and complete the examples below.

Example	Notes
1. Factor $2x^2 + 5x + 3$. List the different factorizations of $2x^2$ and 3. List the possible factoring combinations and check the middle term of each combination. Answer:	
2. Factor $4x^2 - 21x + 5$. List the different factorizations of $4x^2$ and 5. Answer:	

Example	Notes
3. Factor $10x^2 - 9x - 9$.	
Answer:	
4. Factor $3x^2 + 5x - 12$.	
Answer:	

Helpful Hints

When factoring trinomials of the form $ax^2 + bx + c$, there can be many combinations of the factors of ax^2 and c. If the trinomial can be factored, only one of these combinations gives the correct middle term.

If c is positive, the signs of both of its factors are the same as the sign of b. If c is negative, then the signs of both of its factors are different.

Concept Check

1. Does the trinomial $8x^2 + 26x + 15$ factor to $(4x + 5)(2x + 3)$? How can you tell without multiplying the binomials completely?

Practice

Factor.

2. $3x^2 + 7x + 2$

4. $9x^2 + 26x + 16$

3. $10x^2 - 37x + 7$

5. $-x^2 - 5x + 24$

Factoring Polynomials
Topic 14.5 More Factoring of Trinomials

Vocabulary
greatest common factor   •   reverse FOIL method   •   prime polynomial

1. When factoring a trinomial of the form $ax^2 + bx + c$, if c does not have any factors that have a sum of b, then the trinomial cannot be factored and is called a

   _____.

Step-by-Step Video Notes
Watch the Step-by-Step Video lesson and complete the examples below.

Example	Notes
1. Factor $9x^2 + 3x - 30$. Factor out the GCF, then factor the trinomial. Answer:	
2. Factor $3 - 10x + 8x^2$. Answer:	

Example	Notes
3. Factor $5x^2 + 7x + 4$. Answer:	
4. Factor $2x^2 + 12x + 24$. Answer:	

Helpful Hints

When factoring trinomials of the form $ax^2 + bx + c$, always look first for a greatest common factor (GCF). This may be the only possible factoring that can be done. Do not forget to include the GCF in your final answer.

Concept Check

1. Find a whole number value of b between 10 and 20 that would make the trinomial $3x^2 + bx + 16$ a prime polynomial.

Practice

Factor.

2. $8x^2 + 8x - 30$

3. $42x^3 - 45x^2 + 12x$

4. $42x + 20x^2 + 2x^3$

5. $7x^3 + 21x^2 - 14x$

More Factoring and Quadratic Equations
Topic 15.1 Special Cases of Factoring

Vocabulary
difference of two squares   •   perfect square number   •   binomial squared
perfect square trinomial

1. In a _____, the first and last terms are perfect squares, and the middle term is twice the products of square roots of the first and last terms.

Step-by-Step Video Notes
Watch the Step-by-Step Video lesson and complete the examples below.

Example	Notes
2–4. Determine if the expression is a difference of two squares. $x^2 - 16$ $x^2 - 7$ $4x^2 + 81$	
5 & 6. Factor. Remember the property $a^2 - b^2 = (a+b)(a-b)$. $x^2 - 49$ $25b^2 - 64$	

Example	Notes
7 & 8. Factor. $4x^2 - 81y^2$ $-9x^2 + 1$	
9 & 10. Factor the perfect square trinomials completely. $x^2 + 6x + 9$ $9n^2 - 66n + 121$	

Helpful Hints

Other than possibly having a GCF, a sum of two squares will not factor.

If the middle term in a perfect square trinomial is being subtracted, the sign between the terms of the binomial factors will be a minus sign.

Concept Check

1. Can $x^2 - \dfrac{1}{4}$ be factored as a difference of squares? Explain.

Practice

Factor completely.

2. $x^2 - 144$

3. $25x^2 - 81y^2$

4. $16m^2 - 40m + 25$

5. $3x^2 - 42x + 147$

More Factoring and Quadratic Equations
Topic 15.2 Factoring Polynomials

Vocabulary

a difference of two squares   •   perfect square trinomial
greatest common factor   •   reverse FOIL method

1. When factoring a polynomial, always start by looking for a _____.

Step-by-Step Video Notes
Watch the Step-by-Step Video lesson and complete the examples below.

Example	Notes
1. Factor $3k^2 - 48$ completely. Factor out the GCF, if possible. Factor the remaining binomial factor. Answer:	
2. Factor $12n^3 - 12n^2 - 144n$ completely.	

Example	Notes
4. Factor $x^3 + 2x^2 - 9x - 18$ completely.	
5. Factor $4x^3 + 8x^2$ completely.	

Helpful Hints

When factoring polynomials, check for special cases and use different strategies depending on how many terms are in the polynomial. The polynomial will be factored completely when each factor is a prime polynomial.

Concept Check

1. If a polynomial has four terms and no GCF, how should you try to factor?

Practice

Factor completely.

2. $9x^6 - 48x^3 + 64$

4. $64n^8 - 4$

3. $-3x^3 + 18x^2 + 48x - 288$

5. $15x^2 - 23x + 4$

More Factoring and Quadratic Equations
Topic 15.3 Solving Quadratic Equations by Factoring

Vocabulary
Zero Property of Multiplication   •   quadratic equation   •   standard form

1. A _____ is an equation of the form $ax^2 + bx + c = 0$ where $a \neq 0$.

Step-by-Step Video Notes
Watch the Step-by-Step Video lesson and complete the examples below.

Example	Notes
1. Solve $x^2 + 4x = 0$ by factoring. Factor completely. Set each factor equal to zero. Solve each equation. Answer:	
3. Solve $10x^2 - x = 2$ by factoring. Answer:	

Example	Notes
5. Solve $4x^2 + 9 = 12x$ by factoring.	
6. Solve $x^2 - 64 = 0$ by factoring.	

Helpful Hints

The highest degree of any term in a quadratic equation is 2.

The solutions to quadratic equations are also called roots or zeros. A quadratic equation has at most two solutions. Sometimes both solutions will be the same number.

Concept Check

1. Write the quadratic equation $25x^2 + 34x = 4x - 9$ in standard form.

Practice

Solve by factoring.

2. $7x^2 - 28x = 0$

4. $24x^2 + 2x = 35$

3. $3x^2 - 3x - 126 = 0$

5. $25x^2 = 80x - 64$

More Factoring and Quadratic Equations
Topic 15.4 Applications

Vocabulary
quadratic equation   •   roots of a quadratic equation

1. When using a _____ to solve for real-world measurements such as time, distance, length, etc., only use the positive roots of the equation

Step-by-Step Video Notes
Watch the Step-by-Step Video lesson and complete the examples below.

Example	Notes
1. The cliff diver jumps from a platform placed on a cliff approximately 144 feet above the surface of the sea. Disregarding air resistance, the height S, in feet, of a cliff diver above the ocean after t seconds is given by the quadratic equation $S = -16t^2 + 144$. How long does it take the diver to reach the water? (Note: The height when he hits the water is 0 feet.) Answer:	
2. A tennis ball is thrown upward with an initial velocity of 8 meters/second. Suppose that the initial height above the ground is 4 meters. Find the height S of the ball after 1 second. At what time t will the ball hit the ground? Remember, the equation is $S = -5t^2 + vt + h$. Answer:	

Example	Notes
4. The length of the base of a rectangle is 7 inches greater than the height. If the total area of the rectangle is 120 square inches, what are the length of the base and height of the rectangle? For a rectangle, Area $=$ base $\cdot$ height.	
Answer:	

Helpful Hints

The process of solving applications involving quadratic equations is the same as solving quadratic equations in general, with the exception that sometimes there are application specific questions that need to be answered based on the solutions.

Concept Check

1. Write the quadratic equation $3200 = -16t^2 + 480t$ in standard form with a positive coefficient for t^2.

Practice

Solve.

2. An egg is thrown upward with an initial velocity (v) of 9 meters/second. Suppose that the initial height (h) above the ground is 2 meters. At what time t will the egg hit the ground? Use the quadratic equation $S = -5t^2 + vt + h$.

3. A rocket is fired upwards with a velocity (v) of 640 feet per second. Find how many seconds it takes for the rocket to reach a height of 6,400 feet. Use the quadratic equation $S = -16t^2 + 640t$.

4. The length of the base of a rectangle is 4 inches more than twice the height. If the total area of the rectangle is 126 square inches, what are the lengths of the base and height of the rectangle? Remember that for a rectangle, Area $=$ base $\cdot$ height.

Name: _____   Date: _____

Instructor: _____   Section: _____

Introduction to Rational Expressions
Topic 16.1 Undefined Rational Expressions

Vocabulary
rational number • rational expression • zero property of multiplication

1. Any _____ can be written as a fraction of two algebraic expressions.

2. Any _____ can be written as a fraction of two integers.

Step-by-Step Video Notes
Watch the Step-by-Step Video lesson and complete the examples below.

Example	Notes
1. Find any values of the variable that will make the rational expression $\dfrac{15x^2 + 25x}{5x}$ undefined. Set the denominator equal to zero. Solve. Answer:	
2. Find the value of the variable that will make the rational expression $\dfrac{24x^2 + 9x}{8x + 3}$ undefined. Answer:	

Example	Notes
4. Find any values of the variable that will make the rational expression $\dfrac{x^2+6x+8}{x^2-16}$ undefined. Answer:	
5. Find any values that will make the rational expression $\dfrac{x^3+2x^2+x+2}{x^2+1}$ undefined. Answer:	

Helpful Hints

Division by zero is undefined, so the denominator of a rational expression cannot be zero. Any value of the variable that would make the denominator zero is not allowed. Thus, the domain of a rational expression is all values that do not give a zero in the denominator.

Concept Check

1. Is the expression $\dfrac{2x}{x}$ defined for all values of x? Explain.

Practice

Find any values of the variable that will make the rational expression undefined.

2. $\dfrac{68x^3+52x^2}{17x}$

4. $\dfrac{x^2-8}{x^2-12x+32}$

3. $\dfrac{49x^3+56x}{5x-8}$

5. $\dfrac{x^4-25}{x^2+9}$

Introduction to Rational Expressions
Topic 16.2 Simplifying Rational Expressions

Vocabulary
Basic Rule of Fractions   •   simplifying rational expressions   •   common factors

1. The _____ states that for any rational expression $\dfrac{ac}{bc}$ and any

    expressions a, b, and c, (where $b \neq 0$ and $c \neq 0$), $\dfrac{ac}{bc} = \dfrac{a}{b}$.

Step-by-Step Video Notes
Watch the Step-by-Step Video lesson and complete the examples below.

Example	Notes
1. Simplify $\dfrac{21}{39}$. Answer:	
2. Simplify $\dfrac{2x+6}{3x+9}$. Factor the numerator and denominator completely, and divide by common factors. Answer :	

Example	Notes
3. Simplify $\dfrac{x^2+9x+14}{x^2-4}$. Answer :	
5. Simplify $\dfrac{5x^2-45}{45-15x}$. Answer :	

Helpful Hints

Factor the numerator and denominator completely, being aware of special factoring cases like differences of two squares, trinomial squares, monomial factors, and negative factors.

For all polynomials A and B, where $A \neq B$, it is true that $\dfrac{A-B}{B-A}=-1$.

Concept Check

1. What makes the expression $\dfrac{125x^3-9y^2}{9y^2-125x^3}$ easy to simplify? What is the simplified form?

Practice

Simplify.

2. $\dfrac{9x+27}{4x+12}$

4. $\dfrac{x^2+9xy+18y^2}{5x^2+17xy+6y^2}$

3. $\dfrac{5x-4}{12-15x}$

5. $\dfrac{21-4x-x^2}{4x^2-36}$

Introduction to Rational Expressions
Topic 16.3 Multiplying Rational Expressions

Vocabulary
multiplying fractions   •   multiplying rational expressions   •   the quotient rule

1. When multiplying rational expressions, use _____ to simplify the variable parts.

Step-by-Step Video Notes
Watch the Step-by-Step Video lesson and complete the examples below.

Example	Notes
1. Multiply $\dfrac{12}{7} \cdot \dfrac{49}{36}$. Answer:	
2. Multiply $\dfrac{2x^2}{3y} \cdot \dfrac{6y}{8x}$. Factor first. Find common factors in the numerators and the denominators. Divide the numerical parts by common factors, and use the quotient rule to simplify the variable parts. Answer:	

Example	Notes
3. Multiply $\dfrac{7x}{22} \cdot \dfrac{11x+33}{7x+21}$. Answer:	
5. Multiply $\dfrac{x^2-x-12}{16-x^2} \cdot \dfrac{2x^2+7x-4}{x^2-4x-21}$. Answer:	

Helpful Hints

To multiply two rational expressions, find the common factors in the numerators and the denominators. Divide the numerators and denominators by common factors. Then multiply the remaining factors.

Concept Check

1. Are there any common factors to divide in the multiplication $\dfrac{4}{x} \cdot \dfrac{x+4}{4x^2+x}$?

Practice
Multiply.

2. $\dfrac{49x^2}{42y^3} \cdot \dfrac{48y^6}{35x}$

4. $\dfrac{x^4-16}{8x^2+32} \cdot \dfrac{32x^2+24x}{3x^3-4x^2-4x}$

3. $\dfrac{3x-24}{6x+75} \cdot \dfrac{4x+50}{12x-96}$

5. $\dfrac{7-x}{x^2-4x-21} \cdot \dfrac{10-x-x^2}{9x-18}$

Introduction to Rational Expressions
Topic 16.4 Dividing Rational Expressions

Vocabulary
dividing fractions   •   dividing rational expressions   •   reciprocal

1. When _____ multiply the first rational expression by the reciprocal of
    the second rational expression.

Step-by-Step Video Notes
Watch the Step-by-Step Video lesson and complete the examples below.

Example	Notes
2. Divide $\dfrac{-12x^2}{5y} \div \dfrac{18x}{15y}$. Find the reciprocal of the second rational expression and multiply. Divide the numerators and denominators by common factors and then write the remaining factors as one fraction. Answer:	
3. Divide $\dfrac{8x}{14} \div \dfrac{8x-32}{7x-28}$. Answer:	

Example	Notes
4. Divide $\dfrac{x^2+3x-10}{x^2+x-20} \div \dfrac{x^2+4x+3}{x^2-3x-4}$. Answer:	
5. Divide $\dfrac{x-5}{3} \div \left(25-x^2\right)$. Answer:	

Helpful Hints

The reciprocal of an integer or an expression is 1 over the integer or the expression. To get the reciprocal of a fraction or a rational expression, invert the fraction or expression.

Remember when multiplying or factoring that $\dfrac{(a-b)}{(b-a)} = -1$.

Concept Check

1. Dividing by $\dfrac{4x}{3+y}$ is the same as multiplying by what rational expression?

Practice
Divide.

2. $\dfrac{8x^4}{45y^3} \div \dfrac{32x^2}{9y^2}$

4. $\dfrac{14x^2+13x+3}{28x^2+5x-3} \div \dfrac{6x^2-7x-5}{12x^2+17x-5}$

3. $\dfrac{11x}{42} \div \dfrac{11x-77}{6x-42}$

5. $\dfrac{3x-5}{6} \div \left(25-9x^2\right)$

Adding and Subtracting Rational Expressions
Topic 17.1 Adding Like Rational Expressions

Vocabulary
like rational expressions   •   unlike rational expressions   •   numerator

1. Rational expressions with a common denominator are called _____.

Step-by-Step Video Notes
Watch the Step-by-Step Video lesson and complete the examples below.

Example	Notes
1. Add $\dfrac{5a}{4a+2b}+\dfrac{6a}{4a+2b}$. Add the numerators and keep the denominator the same. Answer:	
2. Add $\dfrac{-7m}{2n}+\dfrac{m}{2n}$. Answer:	
3. Add $\dfrac{-3}{x^2-3x+2}+\dfrac{x+1}{x^2-3x+2}$. Answer:	

Example	Notes
4 & 5. Add. $$\frac{x+3}{x^2-1}+\frac{x}{x^2-1}$$ $$\frac{x}{x+1}+\frac{1}{x+1}$$	

Helpful Hints

If rational expressions have a common denominator, they can be added in the same way as like fractions. For any rational expressions $\frac{a}{b}$ and $\frac{c}{b}$, $\frac{a}{b}+\frac{c}{b}=\frac{a+c}{b}$, where $b \neq 0$.

With all calculations with rational expressions, remember to simplify whenever possible by combining like terms, factoring, and dividing by common factors.

Concept Check

1. Are $\dfrac{5}{4x-7}$ and $\dfrac{-3}{(-7)+4x}$ like rational expression?

Practice
Add.

2. $\dfrac{5}{x-8}+\dfrac{3}{x-8}$

3. $\dfrac{3u}{20q}+\dfrac{2u}{20q}$

4. $\dfrac{3x^2-4x}{3x-7}+\dfrac{5x-14}{3x-7}$

5. $\dfrac{4}{x+5}+\dfrac{x+1}{x+5}$

Adding and Subtracting Rational Expressions
Topic 17.2 Subtracting Like Rational Expressions

Vocabulary
 rational expression   •   like rational expression   •   common denominator

1. A _____ can be written as a fraction of two algebraic expressions.

Step-by-Step Video Notes
Watch the Step-by-Step Video lesson and complete the examples below.

Example	Notes
1. Subtract $\dfrac{-2a}{3b} - \dfrac{5a}{3b}$. Subtract the numerators and keep the denominator the same. Answer:	
2. Subtract $\dfrac{8x}{2x+3y} - \dfrac{3x}{2x+3y}$. Subtract the numerators and keep the denominator the same. Answer:	
3. Subtract $\dfrac{3x^2+2x}{x^2-1} - \dfrac{10x-5}{x^2-1}$. Answer:	

Example	Notes
4 & 5. Subtract.	

$$\frac{3x}{x-4} - \frac{12}{x-4}$$

$$\frac{3x}{x^2+3x+2} - \frac{2x-8}{x^2+3x+2}$$

Helpful Hints

If rational expressions have the same denominator, they can be subtracted in the same way as fractions. For any rational expressions $\frac{a}{b}$ and $\frac{c}{b}$, $\frac{a}{b} - \frac{c}{b} = \frac{a-c}{b}$, where $b \neq 0$.

The numerator of the fraction being subtracted must be treated as a single quantity. Use parentheses when subtracting and be careful to use the correct signs.

Concept Check

1. Sophie subtracted $\frac{2x}{2x-3} - \frac{-3}{2x-3}$ and got an answer of 1. Is she correct? Explain.

Practice

Subtract.

2. $\dfrac{7}{5x-2} - \dfrac{8}{5x-2}$

4. $\dfrac{-7a}{4b} - \dfrac{a}{4b}$

3. $\dfrac{9m}{3m+n} - \dfrac{5m+7}{3m+n}$

5. $\dfrac{3x^2+17x}{9x+3} - \dfrac{x-5}{9x+3}$

Adding and Subtracting Rational Expressions
Topic 17.3 Finding the Least Common Denominator for Rational Expressions

Vocabulary
least common denominator (LCD)   •   like rational expression   •   factor

1. The _____ of two or more rational expressions is the smallest expression that each of the denominators will divide into exactly.

Step-by-Step Video Notes
Watch the Step-by-Step Video lesson and complete the examples below.

Example	Notes
1. Find the LCD for $\dfrac{5}{2x-4}$, $\dfrac{6}{3x-6}$. Factor each denominator. List each different factor. List each factor the greatest number of times it occurs in each denominator. Answer:	
2. Find the LCD for $\dfrac{5}{12ab^2c}$, $\dfrac{13}{18a^3bc^4}$. Answer:	

Example	Notes
3–5. Find the LCD. $\dfrac{5}{x+3}, \dfrac{2}{x-4}$ $\dfrac{8}{x^2-5x+4}, \dfrac{12}{x^2+2x-3}$ $\dfrac{x+3}{x^2-6x+9}, \dfrac{10}{2x^2-4x-6}, \dfrac{x}{2}$	

Helpful Hints

If a factor occurs more than once in any one denominator, the LCD will contain that factor repeated the greatest number of times that it occurs in any one denominator.

Be careful when lining up common factors. For example, x and $x-2$ are not common factors, but x and x^2 involve the same factor x, with the highest degree of 2.

Concept Check

1. Is $4x^4$ a common denominator of $\dfrac{5}{2x^3}$ and $\dfrac{y}{2x}$? Is it the LCD of $\dfrac{5}{2x^3}$ and $\dfrac{y}{2x}$? Explain.

Practice
Find the LCD.

2. $\dfrac{5}{9x+24}, \dfrac{11}{21x+56}$

4. $\dfrac{2}{x-3}, \dfrac{7}{x+6}$

3. $\dfrac{13}{30x^2y^3z}, \dfrac{16}{45x^3yz^4}$

5. $\dfrac{x-6}{x^3-4x^2+4x}, \dfrac{9x}{7x^2-21x+14}, \dfrac{4}{x}$

Adding and Subtracting Rational Expressions
Topic 17.4 Adding and Subtracting Rational Expressions

Vocabulary
equivalent rational expression   •   unlike rational expression   •   like rational expressions

1. To add or subtract _____, rewrite each rational expression as a(n)
    _____ whose denominator is the least common denominator.

Step-by-Step Video Notes
Watch the Step-by-Step Video lesson and complete the examples below.

Example	Notes
2. Add $\dfrac{5}{xy} + \dfrac{2}{y}$. Answer:	
4. Add $\dfrac{4y}{y^2 + 4y + 3} + \dfrac{2}{y+1}$. Find the LCD. Rewrite each rational expression with the LCD as the denominator. Add the numerators and keep the denominator the same. Simplify if possible. Answer:	

Example	Notes
5. Subtract $\dfrac{3x-4}{x-2}-\dfrac{5x-6}{2x-4}$. Answer:	
6. Subtract $\dfrac{-3}{x^2+8x+15}-\dfrac{1}{2x^2+7x+3}$. Answer:	

Helpful Hints

It can be very easy to make a sign mistake when subtracting two rational expressions. You will find it helpful to place parentheses around the numerator of the second fraction so that you will not forget to subtract the entire numerator.

Remember that you can only add or subtract rational expressions with like denominators.

Concept Check

1. If $a \cdot b = c$, explain the steps you would use to add $\dfrac{x}{a}+\dfrac{x}{c}$.

Practice

Add.

2. $\dfrac{9}{m}+\dfrac{4}{mn}$

Subtract.

4. $\dfrac{3x+9}{2x-10}-\dfrac{x-3}{x-5}$

3. $\dfrac{3a-b}{a^2-9b^2}+\dfrac{4}{a+3b}$

5. $\dfrac{x}{x^2+3x-4}-\dfrac{x}{x^2+6x+8}$

Complex Rational Expressions and Rational Equations
Topic 18.1 Simplifying Complex Rational Expressions by Adding and Subtracting

Vocabulary
complex rational expressions   •   least common denominator

1. A _____ (also called a complex fraction) is a rational expression that contains a fraction in the numerator, in the denominator, or both.

Step-by-Step Video Notes
Watch the Step-by-Step Video lesson and complete the examples below.

Example	Notes
1. Simplify. $$\frac{\dfrac{1}{x}}{\dfrac{2}{y^2}+\dfrac{1}{y}}$$ Simplify the denominator into a single fraction. Divide the fraction in the numerator by the fraction in the denominator. Answer:	
2. Simplify. $$\frac{\dfrac{1}{x}+\dfrac{1}{y}}{\dfrac{3}{x}-\dfrac{2}{y}}$$ Answer:	

Example	Notes
4. Simplify. $$\dfrac{\dfrac{3}{a+b}-\dfrac{3}{a-b}}{\dfrac{5}{a^2-b^2}}$$ Answer:	

Helpful Hints

The fraction bar in a complex fraction is both a grouping symbol and a symbol for division.

To simplify a complex rational expression by adding and subtracting, simplify the numerator and the denominator into single fractions as necessary. Divide the fraction in the numerator by the fraction in the denominator.

Concept Check

1. Write a multiplication problem that is equivalent to the complex fraction 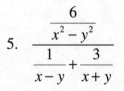 .

Practice
Simplify.

2. $\dfrac{\dfrac{1}{a}+\dfrac{1}{a^2}}{\dfrac{2}{b^2}}$

4. $\dfrac{\dfrac{x}{x^2+4x+3}+\dfrac{2}{x+1}}{x+1}$

3. $\dfrac{\dfrac{1}{x}+\dfrac{1}{y}}{\dfrac{x}{2}-\dfrac{5}{y}}$

5. $\dfrac{\dfrac{6}{x^2-y^2}}{\dfrac{1}{x-y}+\dfrac{3}{x+y}}$

Complex Rational Expressions and Rational Equations
Topic 18.2 Simplifying Complex Rational Expressions by Multiplying by the LCD

Vocabulary
complex rational expressions   •   least common denominator (LCD)

1. One way to simplify a complex fraction is to find the _____ of each denominator in the complex fraction and multiply it by both the numerator and the denominator of the complex fraction.

Step-by-Step Video Notes
Watch the Step-by-Step Video lesson and complete the examples below.

Example	Notes
1. Simplify by multiplying by the LCD. $$\dfrac{\dfrac{3}{x}}{\dfrac{2}{x^2}+\dfrac{5}{x}}$$ Determine the LCD. Multiply the numerator and denominator by the LCD. Answer:	
2. Simplify by multiplying by the LCD. $$\dfrac{\dfrac{5}{ab^2}-\dfrac{2}{ab}}{3-\dfrac{5}{2a^2b}}$$ Answer:	

Example	Notes
3. Simplify by multiplying by the LCD. $$\dfrac{\dfrac{3}{a+b}-\dfrac{3}{a-b}}{\dfrac{5}{a^2-b^2}}$$ Answer:	

Helpful Hints

To simplify a complex rational expression by multiplying by the LCD, you will often have to factor the denominators of the fractions in the expression to determine the LCD.

Concept Check

1. When is it easier to simplify a complex fraction by multiplying by the LCD?

Practice

Simplify by multiplying by the LCD.

2. $\dfrac{\dfrac{3}{a}+\dfrac{2}{b}}{\dfrac{5}{ab}}$

4. $\dfrac{\dfrac{8}{x^2-y^2}}{\dfrac{3}{x+y}+\dfrac{4}{x-y}}$

3. $\dfrac{\dfrac{3}{4x^2}-\dfrac{2}{y}}{\dfrac{7}{xy}-6}$

5. $\dfrac{\dfrac{2x}{x+3}+\dfrac{12}{x^2+8x+15}}{\dfrac{3}{x+5}}$

Complex Rational Expressions and Rational Equations
Topic 18.3 Solving Rational Equations

Vocabulary

rational expression   •   extraneous solution   •   apparent solution

1. A(n) _____ is an apparent solution that does not satisfy the original equation.

Step-by-Step Video Notes
Watch the Step-by-Step Video lesson and complete the examples below.

Example	Notes
1. Solve for x. $$\frac{5}{x} + \frac{2}{3} = -\frac{3}{x}$$ Multiply each term of the equation by the LCD and solve the resulting equation. Answer:	
2. Solve for x. $$\frac{6}{x+3} = \frac{3}{x}$$ Answer:	

Example	Notes
3. Solve for x. $$\frac{3}{x+5} - 1 = \frac{4-x}{2x+10}$$ Answer:	
4. Solve for y. $$\frac{y}{y-2} - 4 = \frac{2}{y-2}$$ Answer:	

Helpful Hints

In the case where a value makes a denominator in the equation zero, it is not a solution to the equation and therefore is not included in the domain.

Concept Check

1. Is 3 an extraneous solution to the equation $\dfrac{x}{x-3} + \dfrac{4}{5} = \dfrac{3}{x-3}$?

Practice

Solve for x.

2. $\dfrac{8}{x} + \dfrac{1}{2} = -\dfrac{2}{x}$

4. $\dfrac{2}{x-5} + 1 = \dfrac{3x-5}{4x-20}$

3. $\dfrac{9}{7x-4} = \dfrac{3}{2x}$

5. $\dfrac{2x}{x-3} = \dfrac{6}{x-3} + 3$

Name: _____   Date: _____

Instructor: _____   Section: _____

Complex Rational Expressions and Rational Equations
Topic 18.4 Direct Variation

Vocabulary
direct variation   •   constant of variation   •   function

1. If y varies directly as x, or y is directly proportional to x, then there is a positive
    constant k such that $y = kx$. In this case, the equation $y = kx$ is called an equation of
    _____.

2. The number k in the direct variation equation $y = kx$ is called the _____
    or the constant of proportionality.

Step-by-Step Video Notes
Watch the Step-by-Step Video lesson and complete the examples below.

Example	Notes
1 & 2. Suppose y varies directly as x. Find the constant of variation and the equation of direct variation for the following. $y = 12$ when $x = 3$ $y = -10$ when $x = -2$	

Example	Notes
3 & 4. Find the variation equation in which y varies directly as x and the following are true. Then find y for the given value of x. $y = -6$ when $x = -3$; $x = 10$ $y = 15$ when $x = 20$; $x = 8$	

Helpful Hints

In direct variation, as x increases or decreases, y also increases or decreases, respectively.

Concept Check

1. Suppose y varies directly as x, and $y = 8$ when $x = 2$. Is it possible to find value of x that corresponds to $y = 12$?

Practice

Suppose y varies directly as x. Find the constant of variation and the equation of direct variation for the following for x.

2. $y = 28$ when $x = 7$

Find the variation equation in which y varies directly as x and the following are true. Then find y for the given value of x.

4. $y = -42$ when $x = -6$; $x = 7$

3. $y = -24$ when $x = -8$

5. $y = 9$ when $x = 12$; $x = 8$

Name: _____   Date: _____

Instructor: _____   Section: _____

Complex Rational Expressions and Rational Equations
Topic 18.5 Inverse Variation

Vocabulary
inverse variation   •   constant of variation   •   function

1. If y varies inversely as x, or y is inversely proportional to x, then there is a positive constant k such that $y = \dfrac{k}{x}$. In this case, the equation $y = \dfrac{k}{x}$ is called an equation of

_____.

2. The number k in the inverse variation equation $y = \dfrac{k}{x}$ is called the

_____ or the constant of proportionality.

Step-by-Step Video Notes
Watch the Step-by-Step Video lesson and complete the examples below.

Example	Notes
1 & 2. Find an equation of variation in which y varies inversely as x for each of the following. $y = 5$ when $x = 2$ $y = -4$ when $x = -7$	

Example	Notes
3 & 4. Find an equation of variation in which y varies inversely as x and the following are true. Then find y for the given value of x. $y = -8$ when $x = -7$; $x = 4$ $y = 18$ when $x = \dfrac{1}{2}$; $x = 27$	

Helpful Hints

In inverse variation, as x increases or decreases, y decreases or increases, respectively.

Concept Check

1. Suppose y varies inversely as x, and $y = 3$ when $x = 1$. Can you find x given that $y = 1$?

Practice

Find an equation of variation in which y varies inversely as x for each of the following.

2. $y = 3$ when $x = 12$

Find an equation of variation in which y varies inversely as x and the following are true. Then find y for the given value of x.

4. $y = -12$ when $x = -4$; $x = -6$

3. $y = -11$ when $x = -2$

5. $y = 8$ when $x = 2.5$; $x = -2$

Complex Rational Expressions and Rational Equations
Topic 18.6 Applications

Vocabulary
inverse variation   •   constant of variation   •   direct variation   •   proportionality

1. The number k in the variation equations $y = kx$ and $y = \dfrac{k}{x}$ is called the

    _____ or the constant of proportionality.

Step-by-Step Video Notes
Watch the Step-by-Step Video lesson and complete the examples below.

Example	Notes
1. The cost, C, of filling a tank of gas varies directly with the number of gallons of gas. If the cost of 1 gallon of gas is $\$2.50$, find the cost of 10 gallons of gas. Let N represent the number of gallons of gas. Answer:	
2. The distance a spring stretches, d, varies directly with the weight of the object hung on the spring, w. If a 10-pound weight stretches a spring 6 inches, how far will a 35-pound weight stretch this spring? Answer:	

Example	Notes
4. If the voltage in an electric circuit is kept at the same level, the current I varies inversely with the resistance, R. The current measures 40 amperes when the resistance is 300 ohms. Find the current when the resistance is 100 ohms.	

Answer:

Helpful Hints
There are many applications that use direct and inverse variation. You will not always be told in a problem statement which kind of variation equation to use to solve a problem.

Concept Check
1. Does the amount of time it takes to drive to a certain destination vary directly or inversely with the average rate of speed you drive?

Practice
2. The amount of money you earn, S, varies directly with the number of hours that you work, t. If you earn $72 when you work 8 hours, determine how much you will earn if you work 13 hours.

3. The cost, C, of producing a certain tool varies inversely as the number produced, n. If 1000 of these tools are produced, the cost is $8 per unit. Find the cost per unit to produce 1600 tools.

4. If the temperature of a gas in a container is constant, the pressure P of the gas varies inversely with the volume V of the container. The pressure is 24 pounds per ft^2 when the volume is 4 ft^3. Find the pressure in a container (of the same temperature) with a volume of 6 ft^3.

Roots and Radicals
Topic 19.1 Square Roots

Vocabulary

square root   •   principal square root   •   negative square root

1. The _____ is one of two identical factors of a number.

2. The radical symbol $\sqrt{}$ is used to denote the _____ of a number.

Step-by-Step Video Notes
Watch the Step-by-Step Video lesson and complete the examples below.

Example	Notes
1–4. Find the square roots.	
$\sqrt{100}$	
$\sqrt{\dfrac{9}{25}}$	
$-\sqrt{49}$	
$\sqrt{0}$	
7 & 8. Find the square roots.	
$\sqrt{49x^2}$	
$\sqrt{(-6x)^2}$	

Example	Notes
9 & 10. Find the square roots. $\sqrt{-144}$ $-\sqrt{-9}$	

---

12. Approximate $\sqrt{75}$ by finding the two consecutive whole numbers that the square root lies between.

Answer:

Helpful Hints

If a variable appears in the radicand, assume it represents positive numbers only.

The square root of a negative number is not a real number.

Not every positive number has a rational square root. You can use a calculator to approximate the square roots of such numbers.

Concept Check

1. The value of $\sqrt{22}$ is between what two whole numbers?

Practice

Find the square roots.

2. $\sqrt{144}$

4. $\sqrt{64z^2}$

3. $-\sqrt{\dfrac{36}{121}}$

5. $\sqrt{-\dfrac{1}{4}}$

Roots and Radicals
Topic 19.2 Higher-Order Roots

Vocabulary

square root   •   cube root   • n^{th} root

1. The _____ of a number is one of three identical factors.

2. The _____ is one of n identical factors of a number.

Step-by-Step Video Notes
Watch the Step-by-Step Video lesson and complete the examples below.

Example	Notes
1–3. Find the cube roots. $\sqrt[3]{27x^3}$ $-\sqrt[3]{\dfrac{8}{27}}$ $\sqrt[3]{-1}$	
4 & 5. Find the indicated roots. $\sqrt[5]{32}$ $\sqrt[4]{-81}$	

Example	Notes
6–8. Simplify the radicals. $\sqrt[5]{x^{10}}$ $\sqrt[4]{y^{12}}$ $\sqrt[4]{a^{24}b^{28}}$	

9 & 10.   Simplify the radicals.

$\sqrt[3]{-64x^6}$

$\sqrt[4]{10,000x^4}$

Helpful Hints

A higher-order root is found using the radical sign $\sqrt[n]{a}$ where n is the index of the radical and a is called the radicand. To find an nth root, we can divide the exponent of the radicand by the index.

The cube root of a negative number is a negative number. If the index is even, the radicand must be nonnegative for the root to be a real number.

Concept Check

1.   Find three different sets of whole number values for n and a if $\sqrt[n]{a} = 2$, and $n \geq 3$.

Practice
Simplify the radicals.

2. $\sqrt[3]{-27}$

3. $\sqrt[6]{x^{30}}$

4.   Fill in the table below.

x	x^2	x^3	x^4	x^5
1	1	1	1	1
2	4		16	
3				243
4			256	
5		125		3125

Name: _____   Date: _____

Instructor: _____   Section: _____

Roots and Radicals
Topic 19.3 Simplifying Radical Expressions

Vocabulary

square root   •   product rule for radicals   •   prime factorization

1. The _____ states that for all nonnegative real numbers a, b, and n, $\sqrt[n]{a} \cdot \sqrt[n]{b} = \sqrt[n]{ab}$ and $\sqrt[n]{ab} = \sqrt[n]{a} \cdot \sqrt[n]{b}$.

Step-by-Step Video Notes
Watch the Step-by-Step Video lesson and complete the examples below.

Example	Notes
1. Simplify the radical $\sqrt{50}$. Factor the radicand. If possible write the radicand as a product of a perfect square. Use the Product Rule to separate the factors. Answer:	
2 & 3. Simplify the radicals. $\sqrt{8}$ $\sqrt{48}$	

Example	Notes
5 & 6. Simplify the radicals. Assume all variables represent nonnegative values. If the answer is not a real number, say so. $$\sqrt{x^5}$$ $$\sqrt[3]{y^{17}}$$	
7 & 8. Simplify the radicals. Assume all variables represent nonnegative values. If the answer is not a real number, say so. $$\sqrt{24x^3}$$ $$\sqrt[3]{-16x^{11}}$$	

Helpful Hints

When simplifying, assume that all variables inside radicals represent nonnegative values.

When simplifying radicals, you can find the perfect root factors in a more difficult problem by looking at the prime factorizations of the radicands.

Concept Check

1. Of $\sqrt[3]{a^2}$, $\sqrt[3]{a^4}$, and $\sqrt[3]{-a^9}$, which expression cannot be simplified?

Practice

Simplify. Assume all variables represent nonnegative values.

2. $\sqrt{75}$

3. $\sqrt{m^7}$

4. $\sqrt[3]{-24}$

5. $\sqrt[3]{-250y^4}$

Roots and Radicals
Topic 19.4 Rational Exponents

Vocabulary
rational exponents   •   the product rule for exponents   •   radical notation

1. _____ can be written as radicals.

2. The _____ states that $x^a \cdot x^b = x^{a+b}$.

Step-by-Step Video Notes
Watch the Step-by-Step Video lesson and complete the examples below.

Example	Notes
1 & 2. Write in radical notation. Simplify, if possible. $9^{1/2}$ $x^{1/3}$	
3 & 4. Write in radical notation. Simplify, if possible. $6^{2/3}$ $x^{3/2}$	

Example	Notes
5 & 6. Use radical notation to rewrite each expression. Simplify if possible. $\left(\dfrac{1}{9}\right)^{3/2}$ $\left(3x\right)^{1/3}$	

9. Use the rules of exponents to simplify.

$$\frac{6^{1/3}}{6^{4/3}}$$

Helpful Hints

If m and n are integers greater than 1, with $\dfrac{m}{n}$ in simplest form, then $a^{m/n} = \sqrt[n]{a^m} = \left(\sqrt[n]{a}\right)^m$,

as long as $\sqrt[n]{a}$ is a real number.

If an exponential expression has a negative rational exponent, write the expression with a positive exponent before evaluating the rational exponent.

Concept Check

1. Express $\sqrt[5]{32x^{10}}$ with rational exponents and simplify if possible.

Practice

Write in radical form. Simplify if possible.   Simplify.

2. $\left(4x\right)^{2/3}$ 4. $x^{2/3} \cdot x^{1/4}$

3. $8^{5/3}$ 5. $16^{-3/2}$

Roots and Radicals
Topic 19.5 The Pythagorean Theorem

Vocabulary
hypotenuse   •   leg   •   the Pythagorean Theorem

1. The side opposite the right angle in a right triangle is called the _____.

2. In a right triangle, _____ states that the sum of the squares of the legs is equal to the square of the hypotenuse, or $leg^2 + leg^2 = hypotenuse^2$.

Step-by-Step Video Notes
Watch the Step-by-Step Video lesson and complete the examples below.

Example	**Notes**
1. Find the length of the hypotenuse. 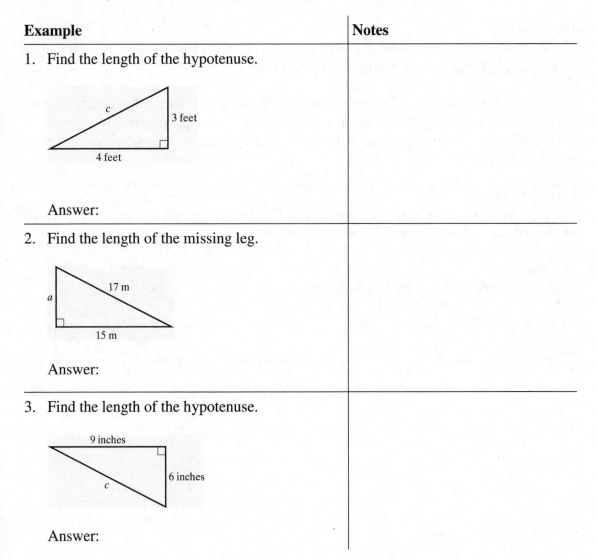 Answer:	
2. Find the length of the missing leg. Answer:	
3. Find the length of the hypotenuse. Answer:	

Example	Notes
4 A slanted roof rises 5 feet vertically from the edge of the roof to the top. The roof covers 12 horizontal feet. How long is the slanted surface of the roof?	

Answer:

Helpful Hints

The Pythagorean Theorem is often presented as $a^2 + b^2 = c^2$, where a and b represent the legs of a right triangle and c represents the hypotenuse.

When using the Pythagorean Theorem, if the missing length is not a rational number, then express it either as a simplified radical or use a calculator to approximate it to a certain number of places. A given problem will usually specify whether an approximated or exact answer is required.

Concept Check

1. Why can't you always use the Pythagorean Theorem to find the missing side of a triangle with a side of length 2 m and a side of length 8 m?

Practice
Find the missing length in each right triangle.

2.   3.   4.

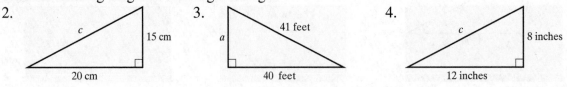

Roots and Radicals
Topic 19.6 The Distance Formula

Vocabulary

the Distance Formula   •   the Pythagorean Theorem

1. To calculate the distance between any two points (x_1, y_1) and (x_2, y_2) on a graph, use

   _____, which states $d = \sqrt{(x_2 - x_1)^2 + (y_2 - y_1)^2}$.

Step-by-Step Video Notes

Watch the Step-by-Step Video lesson and complete the examples below.

Example	Notes
1. Find the length of the line segment between the two points shown. Answer:	
3. Find the distance between $(-3, 2)$ and $(1, 5)$. Answer:	

Example	Notes
4 & 5. Use the Distance Formula to find the distance between the given points. $(6,-3)$ and $(1,9)$ $(2,1)$ and $(6,8)$	
6. Use the Distance Formula to find the distance between the points $(-13,-8)$ and $(-3,-3)$. Answer:	

Helpful Hints

Notice in example 3 that the horizontal and vertical distances between the two points make up the legs of a right triangle. The hypotenuse of this right triangle is the distance between the two points. This is how the Distance Formula is derived from the Pythagorean Theorem.

Be careful to avoid sign errors when finding the differences between the x- and y-coordinates of the points.

Concept Check
1. Will the Distance Formula also work for two points on the same horizontal line or the same vertical line? Explain.

Practice

Find the distance between the two points shown on the graph.
2.

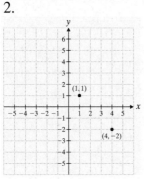

Use the Distance Formula to find the distance between the given points.

3. $(-3,-4)$ and $(5,11)$

4. $(4,7)$ and $(8,13)$

Name: _____   Date: _____

Instructor: _____   Section: _____

Operations of Radical Expressions
Topic 20.1 Adding and Subtracting Radical Expressions

Vocabulary
like radicals   •   like terms   •   radical expressions

1. Radicals with the same radicand and the same index are called _____.

Step-by-Step Video Notes
Watch the Step-by-Step Video lesson and complete the examples below.

Example	Notes
1–3. Combine like terms. $7x + 9x$ $13xy^2 + 11x^2y$ $5x^{1/2} + 2x^{1/2}$	
4–6. Simplify, if possible. Assume that all variables are nonnegative real numbers. $5\sqrt{3} + 7\sqrt{3}$ $3\sqrt{2} - 9\sqrt{2}$ $-6\sqrt{xy} + 2\sqrt[3]{xy}$	

Example	Notes
8. Simplify. $5\sqrt{3} - \sqrt{27} + 2\sqrt{32}$	
10. Simplify. Assume that all variables are nonnegative real numbers. $3x\sqrt[3]{54x^4} - 3\sqrt[3]{16x^7}$	

Helpful Hints

A radical is written using the radical sign $\sqrt[n]{a}$ where n is the index of the radical and a is called the radicand.

For all nonnegative real numbers a, b, and n, where n is an integer greater than 1, $\sqrt[n]{ab} = \sqrt[n]{a} \cdot \sqrt[n]{b}$.

Concept Check

1. Are $x\sqrt{7}$ and $7\sqrt{x}$ like radicals? Explain why or why not.

Practice

Simplify. Assume variables represent nonnegative numbers.

2. $11\sqrt{7} - 12\sqrt{7}$

4. $-\sqrt{12} + 6\sqrt{27} - 4\sqrt{28}$

3. $\sqrt{48} + \sqrt{75}$

5. $\sqrt{72x} - 2x\sqrt{3} - 5\sqrt{2x} + x\sqrt{27}$

Name: _____ Date: _____

Instructor: _____ Section: _____

Operations of Radical Expressions
Topic 20.2 Multiplying Radical Expressions

Vocabulary
Distributive Property   •   Product Rule for Radicals   •   FOIL method

1. The _____ states that for all nonnegative real numbers a, b, and n, where n is an integer greater than 1, $\sqrt[n]{a} \cdot \sqrt[n]{b} = \sqrt[n]{ab}$.

Step-by-Step Video Notes
Watch the Step-by-Step Video lesson and complete the examples below.

Example	Notes
2. Multiply. $$\sqrt{5} \cdot \sqrt{3}$$ Use the Product Rule for radicals. Answer:	
4. Multiply. $$\left(\sqrt{12}\right)\left(-5\sqrt{3}\right)$$ Use the Product Rule for radicals. Simplify, if possible. Answer:	

Example	Notes
6. Multiply. $$\left(\sqrt{2}+3\sqrt{5}\right)\left(2\sqrt{2}-\sqrt{5}\right)$$ Answer:	
8. Simplify. Assume all variables represent nonnegative numbers. $$\left(\sqrt{7}+\sqrt{3x}\right)^{2}$$ Answer:	

Helpful Hints

Apply multiplicative properties like the Distributive Property, the Product Rules, the FOIL method, the difference of two squares and squaring binomials when multiplying radical expressions, just as you would with other types of expressions.

Recall that $\left(\sqrt{x}\right)^{2}=x$.

Concept Check

1. Addison claims that $\left(\sqrt{-3}\right)\left(\sqrt{-3}\right)=\left(\sqrt{9}\right)=3$. Is she correct? Explain why or why not.

Practice

Multiply. Assume all variables represent nonnegative numbers.

2. $\left(4x\sqrt{2y}\right)\left(7x\sqrt{8}\right)$

4. $\left(4\sqrt{5}+3\sqrt{2}\right)\left(4\sqrt{5}-3\sqrt{2}\right)$

3. $\sqrt{3x}\left(2\sqrt{6x}+7\sqrt{12}\right)$

5. $\sqrt[3]{7x}\left(\sqrt[3]{49x^{2}}-5\sqrt[3]{2x}\right)$

Operations of Radical Expressions
Topic 20.3 Dividing Radical Expressions

Vocabulary
Quotient Rule for Radicals   •   radical expressions

1. The _____ states that for all nonnegative real numbers

    a, b, and n, where n is an integer greater than 1, and $b \neq 0$, $\dfrac{\sqrt[n]{a}}{\sqrt[n]{b}} = \sqrt[n]{\dfrac{a}{b}}$.

Step-by-Step Video Notes
Watch the Step-by-Step Video lesson and complete the examples below.

Example	Notes
1. Divide. $$\frac{\sqrt{75}}{\sqrt{3}}$$ Answer:	
2 & 3. Divide. $$\frac{\sqrt{9}}{\sqrt{16}}$$ $$\frac{\sqrt{72}}{\sqrt{8}}$$	

Example	Notes
4 & 5. Divide.	
$$\sqrt[3]{\dfrac{-a^9}{b^6}}$$	
$$\dfrac{\sqrt[3]{x^8}}{\sqrt[3]{x^5}}$$	

7. Divide. Assume variables represent nonnegative values. $$\dfrac{\sqrt{25x^5y^2}}{\sqrt{144x^3y^4}}$$ Answer:	

Helpful Hints

The Quotient Rule for Radicals can be very flexible. You can simplify the numerators and/or denominators first, or you can divide the radicands first, depending on the situation.

Concept Check

1. Simplify Example 5 a different way than you did originally. Do you get the same answer?

Practice

Divide. Assume all variables represent nonnegative numbers.

2. $\dfrac{\sqrt{150}}{\sqrt{6}}$

4. $\dfrac{\sqrt{25a^6}}{\sqrt{81b^{12}c^4}}$

3. $\dfrac{\sqrt{63a^9b^7}}{\sqrt{7a^3b}}$

5. $\sqrt{\dfrac{147a^2b^6}{3a^8b^4}}$

Operations of Radical Expressions
Topic 20.4 Rationalizing the Denominator

Vocabulary
rationalizing the denominator • Identity Property of Multiplication • conjugates

1. Performing operations to remove a radical from a denominator is called

_____.

Step-by-Step Video Notes
Watch the Step-by-Step Video lesson and complete the examples below.

Example	Notes
1. Simplify by rationalizing the denominator. $$\frac{1}{\sqrt{2}}$$ Answer:	
3. Simplify by rationalizing the denominator. $$\sqrt[3]{\frac{2}{3x^2}}$$ Answer:	
4 . Simplify by rationalizing the denominator. $$\frac{5}{3+\sqrt{2}}$$ Answer:	

Example	Notes
5. Simplify by rationalizing the denominator. $$\dfrac{\sqrt{7}+\sqrt{3}}{\sqrt{7}-\sqrt{3}}$$ Answer:	

Helpful Hints

A rational expression is not considered to be in simplest form if there is an irrational expression in the denominator.

When rationalizing a denominator, you can simplify the radical in the denominator first, and then rationalize the denominator, or you can rationalize first, and then simplify the fraction.

The product of two conjugates is always rational.

Concept Check

1. Explain why $\dfrac{2}{1-\sqrt{2}}$ cannot be rationalized by multiplying by $\dfrac{\sqrt{2}}{\sqrt{2}}$.

Practice

Simplify by rationalizing the denominator. Assume variables represent nonnegative numbers.

2. $\dfrac{2x}{\sqrt{7}}$

4. $\sqrt[3]{\dfrac{4}{3x^2}}$

3. $\dfrac{5}{\sqrt{75}}$

5. $\dfrac{8}{3-\sqrt{5}}$

Operations of Radical Expressions
Topic 20.5 Solving Radical Equations

Vocabulary
Squaring Property of Equality   •   radical equations   •   reverse operations

1. The _____ states that for all real numbers a, and b, if $a = b$, then $a^2 = b^2$.

Step-by-Step Video Notes
Watch the Step-by-Step Video lesson and complete the examples below.

Example	Notes
1. Solve. $\sqrt{x} = 7$ Square both sides of the equation. Check. Answer:	
3. Solve $3 + \sqrt{x} = 7$. Perform operations to get the radical by itself on one side of the equation. Square both sides of the equation and solve. Answer:	

Example	Notes
4. Solve. $$\sqrt{x-5}=12$$ Answer:	
5. Solve. $$\sqrt{5x+6}=x$$ Answer:	

Helpful Hints

Just as you can apply the four basic operations to both sides of an equation, squaring both sides of an equation will result in an equivalent equation.

Check all possible solutions to make sure they work in the original equation. Solutions to radical equations must be verified.

Concept Check

1. Is 64 a solution to $-\sqrt{x}=-8$? Is 64 a solution to $\sqrt{-x}=-8$? Explain why or why not for both cases.

Practice
Solve.

2. $-4+\sqrt{x}=5$

4. $\sqrt{3x+4}=x$

3. $\sqrt{2x-5}=3$

5. $\sqrt{x+6}+8=14$

Solving Quadratic Equations
Topic 21.1 Introduction to Solving Quadratic Equations

Vocabulary

quadratic equation   •   solution   •   parabola   • x-intercept

1. A(n) _____ is an equation of the form $ax^2 + bx + c = 0$ where $a \neq 0$.

2. A(n) _____ of the graph of an equation is a point where the curve crosses the x-axis, or where the value of y is 0.

Step-by-Step Video Notes
Watch the Step-by-Step Video lesson and complete the examples below.

Example	**Notes**
1–3. Write each quadratic equation in standard form. $7x^2 = 5x + 8$ $6x - 2x^2 = -3$ $4x^2 = 9$	
4–6. Determine the number of solutions for each quadratic equation graphed below. 	

Example	Notes

7 & 8. Find the real solution(s) of each quadratic equation, if any exist, by using the graph of the equation.

$y = x^2 + 2x - 3$

$y = -x^2 + 4x - 4$

Helpful Hints

The graph of a quadratic equation $y = ax^2 + bx + c$ is called a parabola.

A quadratic equation may have two real solutions, one real solution, or no real solutions.

Concept Check

1. How can you use the graph of a quadratic equation to determine the real solutions of the equation?

Practice

Write each quadratic equation in standard form.

2. $x^2 + 9 = -6x$

3. $2x + 3 = 5x^2$

Find the real solution(s) of each quadratic equation by using the graph of the equation.

4.

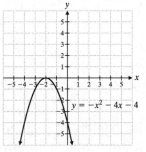

$y = -x^2 - 4x - 4$

5.

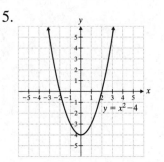

$y = x^2 - 4$

Solving Quadratic Equations
Topic 21.2 Solving Quadratic Equations by Factoring

Vocabulary
standard form   •   factor   •   quadratic expression

1. To solve a quadratic equation by factoring, first make sure the equation is in the
   _____, $ax^2 + bx + c = 0$.

Step-by-Step Video Notes
Watch the Step-by-Step Video lesson and complete the examples below.

Example	Notes
1. Solve $x^2 - 9x = 0$ by factoring. Factor completely. Set each factor equal to zero. Solve each equation. Check each solution. Answer:	
2. Solve $5x^2 - 14x - 3 = 0$ by factoring. Answer:	

Example	Notes
3. Solve $9x^2 = 24x - 15$ by factoring.	
Answer:	
4. Solve $x^2 - 2x = -1$ by factoring.	
Answer:	

Helpful Hints

When solving a quadratic equation by factoring, set each factor containing a variable equal to zero. Remember, if you factor out the variable or a monomial term containing the variable as the GCF, then 0 is one of the solutions of the equation.

Concept Check

1. Rewrite $y = (x - 7)^2$ by setting each factor equal to zero to solve. How many solutions does this equation have?

Practice

Solve by factoring.

2. $3x^2 + 15x = 0$

3. $5x^2 + 3x - 14 = 0$

4. $7x^2 + 84 = 56x$

5. $x^2 + 10x + 25 = 0$

Solving Quadratic Equations
Topic 21.3 Solving Quadratic Equations by Using the Square Root Property

Vocabulary
Square Root Property   •   quadratic equation

1. A _____ is an equation of the form $ax^2 + bx + c = 0$ where $a \neq 0$.

Step-by-Step Video Notes
Watch the Step-by-Step Video lesson and complete the examples below.

Example	Notes
2. Solve $x^2 = 48$ for x. Use the Square Root Property. Simplify and check your solutions. Answer:	
3. Solve $x^2 = -4$ for x. Answer:	

Example	Notes
4. Solve $3x^2 + 2 = 77$ for x.	
Answer:	
5. Solve $(4x-1)^2 = 5$ for x.	
Answer:	

Helpful Hints

The notation $\pm\sqrt{a}$ is a shorthand way of writing "$+\sqrt{a}$ or $-\sqrt{a}$".

The Square Root Property can only be used to solve an equation of the form $x^2 = a$ if $a \geq 0$. If $a < 0$, the equation has no real solutions.

Concept Check

1. Mark uses the square root property to solve the equation $x^2 + 4 = 0$ and incorrectly gets $x = \pm 2$ as his answer. Explain his error and state the correct answer.

Practice

Solve for x.

2. $x^2 = 28$

3. $x^2 = 196$

4. $6x^2 + 2 = 98$

5. $(2x-3)^2 = 25$

Name: _____   Date: _____
Instructor: _____   Section: _____

Solving Quadratic Equations
Topic 21.4 Solving Quadratic Equations by Completing the Square

Vocabulary
completing the square   •   the Square Root Property   •   perfect square trinomial

1. To solve an equation like $x^2 + bx = c$, we can add a constant to both sides of the equation so that the left side becomes a perfect square trinomial. This method is called

_____.

Step-by-Step Video Notes
Watch the Step-by-Step Video lesson and complete the examples below.

Example	Notes
1 & 2. Fill in the blanks to create a perfect square trinomial. $$x^2 + 8x + \underline{} = \left(x + \underline{}\right)^2$$ $$x^2 - 12x + \underline{} = \left(x - \underline{}\right)^2$$	
3. Solve $x^2 + 2x = 3$ by filling in the blanks and then using the Square Root Property. $$x^2 + 2x + \underline{} = 3 + \underline{}$$ Answer:	

Example	Notes
4. Solve $x^2 + 4x - 5 = 0$ by completing the square. Answer:	
5. Solve $x^2 + 6x + 1 = 0$ by completing the square. Answer:	

Helpful Hints

Recall that when a binomial with is squared using FOIL, the coefficient of x is twice the product of coefficient of x in the binomial and the constant of the binomial.

If the equation you are solving contains fractions or if the coefficient of x is odd, then the equation will be more easily solved by using a method other than completing the square.

Concept Check

1. Which method would you use to solve $x^2 - 8x = -9$? Which method would you use to solve $x^2 - 8x = -16$?

Practice

Solve by completing the square.

2. $x^2 + 5x - 6 = 0$

3. $x^2 + 12x + 4 = 0$

4. $2x^2 + 16x - 96 = 0$

5. $x^2 - 4x - 16 = 0$

Solving Quadratic Equations
Topic 21.5 Solving Quadratic Equations by Using the Quadratic Formula

Vocabulary
standard form   •   Square Root Property   •   quadratic formula

1. For all quadratic equations in the form $ax^2 + bx + c = 0$, you can solve by using the

_____, which states that $x = \dfrac{-b \pm \sqrt{b^2 - 4ac}}{2a}$.

Step-by-Step Video Notes
Watch the Step-by-Step Video lesson and complete the examples below.

Example	Notes
1. Solve $3x^2 - x - 2 = 0$ by using the quadratic formula. Identify a, b, and c. Substitute a, b, and c into the quadratic formula. Simplify. Answer:	
2. Solve $x^2 = 6x$ by using the quadratic formula. Answer:	

Example	Notes
3. Solve $4x^2 + 25 = 20x$ by using the quadratic formula.	
Answer:	
5. Solve $x^2 + 4x - 8 = 0$ by using the quadratic formula.	
Answer:	

Helpful Hints

The quadratic formula is the only method of solving that works for every quadratic equation.

Be sure the equation is in standard form before you identify a, b, and c. If b is positive, then $-b$ in the formula is negative, but if b is negative, then $-b$ is positive in the formula.

Sometimes the value of b or c will be equal to 0. If there is no x term, then the value of b is equal to 0; if there is no constant, then the value of c is equal to 0.

Concept Check

1. How many real solutions does a quadratic equation in the form $ax^2 + bx + c = 0$ have if $b^2 - 4ac = 0$? If $b^2 - 4ac < 0$?

Practice

Solve by using the quadratic formula.

2. $x^2 - 8x + 7 = 0$

4. $9x^2 = 6x - 1$

3. $x^2 + 8x = -2 + 3x$

5. $6x = 4x^2 + 3$

Graphing Quadratic Equations
Topic 22.1 Introduction to Graphing Quadratic Equations

Vocabulary
vertex   •   axis of symmetry   •   quadratic equation

1. The _____ of a parabola is the lowest or highest point on a parabola.

2. The _____ of a parabola is the vertical line through the vertex.

Step-by-Step Video Notes
Watch the Step-by-Step Video lesson and complete the examples below.

Example	Notes
1–3. Determine whether the graph of each quadratic equation opens upward or downward. $y = 2x^2 - 6x + 3$ $y = -5x^2 + 11$ $y = -x^2 + 4x - 1$	

Example	Notes
4–6. If an item is dropped from a height of 400 feet, its height above the ground t seconds after being dropped is given by the equation $h = -16t^2 + 400$. Determine the height of the object for the following values of t. 0 seconds 1 second 5 seconds	

Helpful Hints

Graphs of quadratic equations $y = ax^2 + bx + c$ are parabolas opening upward if $a > 0$ or downward if $a < 0$.

The parabola is symmetric over the axis of symmetry, which means that the graph looks the same on both sides.

Concept Check

1. If the vertex of a parabola is the highest point, does the parabola open upward or downward? What is the sign of the leading coefficient?

Practice

Determine whether the graph of each quadratic equation opens upward or downward.

2. $y = -3x^2 + x + 25$

3. $y = \dfrac{2}{3}x^2 + 9x - \dfrac{8}{9}$

Determine the height of an object dropped from a height of 600 feet for each value of t. Use the equation $h = -16t^2 + 600$.

4. $t = 6$ seconds

5. $t = 3$ seconds

Graphing Quadratic Equations
Topic 22.2 Finding the Vertex of a Quadratic Equation

Vocabulary

vertex   •   minimum   •   maximum

1. The _____ of a parabola is the lowest or highest point on a parabola.

Step-by-Step Video Notes
Watch the Step-by-Step Video lesson and complete the examples below.

Example	Notes
1. Find the vertex of the quadratic equation. Is the vertex a maximum or a minimum? $$y = x^2 - 8x + 15$$ Identify a, b, and c, then find $x = \dfrac{-b}{2a}$.	
Substitute this x value into the equation and find for the y value of the vertex.	
Determine if the vertex is a maximum or minimum.	
Answer:	

Example	Notes
2. Find the vertex of the quadratic equation. Is the vertex a maximum or a minimum? $y = 12x - 3x^2 - 6$ Answer:	
3. Find the vertex of the quadratic equation. Is the vertex a maximum or a minimum? $y = 5x^2 - 7$ Answer:	

Helpful Hints

To find the vertex of a quadratic equation, first identify a, b, and c. Then find the x value of the vertex, which is $x = \dfrac{-b}{2a}$. Substitute this x value into the equation and find the y value of the vertex. If $a > 0$ the vertex is a minimum and if $a < 0$ the vertex is a maximum.

Concept Check

1. Will the vertex of the graph of $y + x^2 = -6x + 4$ be a maximum or a minimum?

Practice

Find the vertex of the quadratic equation. Is the vertex a maximum or a minimum?

2. $y = x^2 - 9x + 8$ 　　　　　　　　4. $y = -7x^2 + 4x - 3$

3. $y = -4x^2 + 12x - 9$ 　　　　　　5. $y = 3x^2 - 15$

Graphing Quadratic Equations
Topic 22.3 Finding the Intercepts of a Quadratic Equation

Vocabulary
intercept   • x-intercept   • y-intercept   •   quadratic function

1. To find the _____ of a quadratic function, let $x = 0$ and simplify.

2. To find the _____(s) of a quadratic function (if any exist), solve
    the equation $f(x) = 0$ for x.

Step-by-Step Video Notes
Watch the Step-by-Step Video lesson and complete the examples below.

Example	Notes
1 & 2. Find the y-intercept of each quadratic function. $f(x) = x^2 + 6x + 15$ $f(x) = 3x^2$	
3 & 4. Find the x-intercept(s) of each quadratic function, if any exist. $f(x) = x^2 + 2x - 24$ $f(x) = x^2 + 1$	

Example	Notes
5. Find the x- and y-intercepts of the quadratic function. $$f(x) = x^2 + 5x - 14$$ Answer:	

Helpful Hints

A quadratic function may have zero, one, or two x-intercepts. However, it will always have exactly one y-intercept.

Concept Check

1. Why must a quadratic function have no more than one y-intercept?

Practice

Find the x- and y-intercepts of each quadratic function.

2. $f(x) = x^2 - 6x - 16$

4. $f(x) = x^2 + 11$

3. $f(x) = 3x^2 + 18x + 27$

5. $f(x) = x^2 - 9x + 20$

Graphing Quadratic Equations
Topic 22.4 Graphing Quadratic Equations

Vocabulary
intercepts   •   axis of symmetry   •   parabola   •   vertex

1. A parabola is symmetric over the _____, which means that the graph looks the same on both sides.

Step-by-Step Video Notes
Watch the Step-by-Step Video lesson and complete the examples below.

Example	Notes
1. Graph the equation $f(x) = x^2$. Find and plot ordered pairs that satisfy the function. Draw a smooth curve through the points.	
2. Graph the equation $f(x) = x^2 - 6x + 8$. Determine the direction of the parabola. Find and plot the vertex and the intercepts of the function. Draw a smooth curve through the points.	

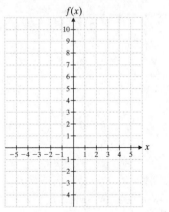

Example	Notes

3. Graph the equation $f(x) = -2x^2 + 4x - 3$.

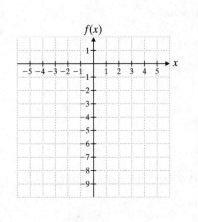

Helpful Hints

When graphing a quadratic equation, determine if the parabola opens upward or downward, making the vertex either a minimum or a maximum, respectively. Identify and plot the vertex, the x-intercept(s) and the y-intercept, and other ordered pairs that satisfy the function, if needed. Draw a smooth curve through the points to form the parabola.

Concept Check

1. For examples 1 through 3, find and graph the axis of symmetry.

Practice

Graph each equation.

2. $f(x) = x^2 - 2x - 3$

3. $g(x) = -2x^2 + 2$

4. $f(x) = -x^2 + 4x - 5$

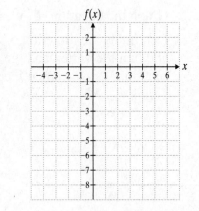

Additional Topics
Topic A.1 The Midpoint Formula

Vocabulary
midpoint   •   midpoint formula   •   length

1. The _____ of a line segment is the point on the segment that is the same distance from each of the endpoints of the line segment.

Step-by-Step Video Notes
Watch the Step-by-Step Video lesson and complete the examples below.

Example	Notes
1. Find the midpoint of the line segment whose endpoints are $(1,4)$ and $(3,2)$. Let (x_1, y_1) represent one point and (x_2, y_2) represent the other. $(\Box, \Box)$ 　x_1　y_1 $(\Box, \Box)$ 　x_2　y_2 Substitute the values into the midpoint formula and simplify. $\dfrac{x_1 + x_2}{2} = \dfrac{\Box + \Box}{2} = \dfrac{\Box}{2} = \Box$ $\dfrac{y_1 + y_2}{2} = \dfrac{\Box + \Box}{2} = \dfrac{\Box}{2} = \Box$ Answer:	

Example	Notes
2–4. Find the midpoint of each line segment whose endpoints are the following points. $(-5,0)$ and $(9,-4)$ $\left(\dfrac{1}{2},-\dfrac{3}{4}\right)$ and $\left(-\dfrac{1}{2},\dfrac{1}{4}\right)$ $(20,-6.5)$ and $(-13,-7.3)$	

Helpful Hints
Do not confuse finding the midpoint of a line segment with finding the distance of a line segment.

The distance between two points is a length, whereas the midpoint of a line segment is a point on the line.

Concept Check
1. What is the relationship between the distance between an endpoint of a segment and the midpoint of a line segment and the distance between the two endpoints of a line segment?

Practice
Find the midpoint of each line segment whose endpoints are the following points.

2. $(2,3)$ and $(4,5)$

4. $\left(\dfrac{2}{3},-\dfrac{2}{3}\right)$ and $\left(-\dfrac{2}{3},\dfrac{1}{3}\right)$

3. $(-7,0)$ and $(9,-6)$

5. $(-2,-3.4)$ and $(15,-7.2)$

Additional Topics
Topic A.2 Writing Equations of Parallel and Perpendicular Lines

Vocabulary
parallel lines   •   perpendicular lines   •   negative reciprocals

1. _____ are lines that are always the same distance apart and have no point of intersection.

2. _____ are lines that intersect at a 90° angle.

Step-by-Step Video Notes
Watch the Step-by-Step Video lesson and complete the examples below.

Example	Notes
1. Find the slope of a line parallel to $y = \dfrac{3}{4}x + 2$. Find the slope of the line. Find the slope of a parallel line. Answer:	
2. Find the slope of a line perpendicular to $6x + 2y = 9$. Find the slope of the line. Find the slope of a perpendicular line. Answer:	

Example	Notes
4. Determine if the lines are parallel, perpendicular, or neither. $y = 2x + 3$ $-8x + 4y = -4$ Answer:	
7. Find the equation of a line perpendicular to $y = \dfrac{3}{5}x + 8$ that passes through the point $(3, -1)$. Write the answer in slope-intercept form. Answer:	

Helpful Hints

The symbol for parallel lines is $\parallel$. The symbol for perpendicular lines is $\perp$.

All horizontal lines are parallel to each other and all vertical lines are parallel to each other. Horizontal lines are perpendicular to vertical lines.

Concept Check

1. Is it possible for a set of two lines to be both parallel and perpendicular? Explain.

Practice

Find the slope of a line parallel to and a line perpendicular to the given line.

2. $5x + 3y = 12$

3. $y = 5x + 3.5$

4. Find the slope of a line containing the points $(-2, 3)$ and $(4, 3)$. Then find the slope of a line parallel to this line and the slope of a line perpendicular to this line.

Additional Topics
Topic A.3 Graphing Linear Inequalities in Two Variables

Vocabulary
linear inequality   •   test point   •   integer solution

1. When graphing a _____, first replace the inequality symbol with an equality symbol. Then graph the line.

Step-by-Step Video Notes
Watch the Step-by-Step Video lesson and complete the examples below.

Example	Notes
1. Graph $4x + 3y \leq 9$. Graph the line. $(\square,\square)\ (\square,\square)\ (\square,\square)$ Test a point. $(\square,\square)$ $4(\square) + 3(\square) \leq 9$ This is a _____ statement. Shade to show all solutions. Answer:	

Example	Notes

2. Graph $3y < -2x$.

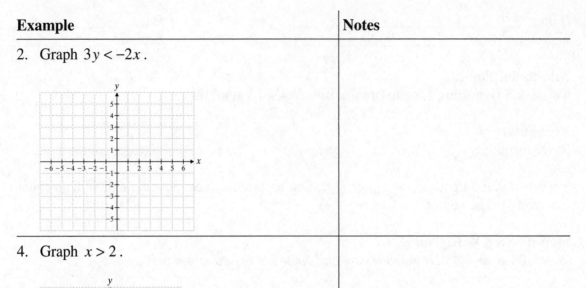

4. Graph $x > 2$.

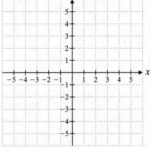

Helpful Hints

The origin $(0,0)$ is usually the most convenient test point to pick. If the point $(0,0)$ is a point on the line, choose another convenient test point.

When the inequality symbol is $\leq$ or $\geq$, use a solid line when graphing the inequality. When the inequality symbol is $<$ or $>$, use a dotted line when graphing the inequality.

Concept Check

1. What test point would you choose to test the linear inequality $y > x$? Explain.

Practice

Graph the given inequality.

2. $y \geq 2x + 3$ 3. $y \leq -2$ 4. $x - y \leq 0$

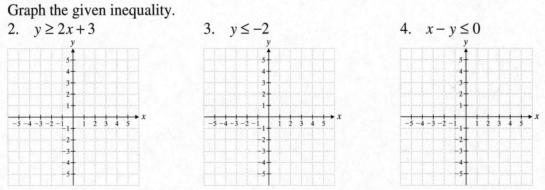

Name: _____   Date: _____

Instructor: _____   Section: _____

Additional Topics
Topic A.4 Systems of Linear Inequalities

Vocabulary
system of linear inequalities   •   test point   •   point of intersection

1. Two or more linear inequalities graphed on the same set of axes is called a

   _____.

Step-by-Step Video Notes
Watch the Step-by-Step Video lesson and complete the examples below.

Example	**Notes**
2. Graph the solution to the system of linear inequalities. $y < -\dfrac{4}{3}x + 3$ $y \geq \dfrac{1}{2}x - 2$ 	
3. Graph the solution to the system of linear inequalities. $y \geq 2x + 3$ $4x - 2y > 4$ 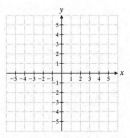	

Example	Notes
4. Graph the solution to the system of linear inequalities. Find and label all the points of intersection. $x + y \le 3$ $x - y \le 1$ $x \ge -1$	

Helpful Hints

The solution to a system of linear inequalities is the intersection of the solution sets of the individual inequalities.

If the graph of a system of linear inequalities is parallel lines with the shading outside the lines, there is no solution to the system.

Concept Check

1. Can a system of two linear inequalities have no solution if it has two perpendicular lines?

Practice

Graph the solution to the given system of inequalities.

2. $y \ge 0.5x$
   $x \ge 2y + 3$

3. $y \ge 3$
   $x < -1$

4. $y - x > 2$
   $y < 2$
   $y \ge -2x - 4$

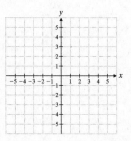

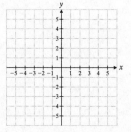

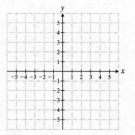

Additional Topics
Topic A.5 Synthetic Division

Vocabulary
synthetic division   •   quotient   •   divisor   •   dividend

1. Dividing a polynomial by a binomial can be made more efficient with a process called _____.

Step-by-Step Video Notes
Watch the Step-by-Step Video lesson and complete the examples below.

Example	Notes
1. Divide using synthetic division. $\left(3x^3 + 7x^2 - 4x + 3\right) \div \left(x + 3\right)$ $\begin{array}{c|cccc} \square & \square & \square & \square & \square \\ & & \square & \square & \square \\ \hline & \square & \square & \square & \square \end{array}$ Turn the first three numbers in the bottom row into the quotient. The last number in the bottom row is the remainder. Put this over the divisor to complete the answer.	

Answer:

Example	Notes
2. Divide using synthetic division. $\left(3x^4 - 21x^3 + 31x^2 - 25\right) \div \left(x - 5\right)$	

Helpful Hints

"Missing terms" refers to monomial terms with degrees between the degrees of the terms in the polynomial being divided. For example, if the polynomial being divided is $3x^3 + 1$, the missing terms are $0x^2$ and $0x$.

Be sure to include zeros for the missing terms when using synthetic division.

Concept Check

1. What is the disadvantage of synthetic division?

Practice

Divide using synthetic division.

2. $\left(x^3 + 4x + 4\right) \div \left(x + 1\right)$

3. $\left(x^5 + 4x^2 + x\right) \div \left(x - 2\right)$

4. $\left(x^4 + 3x^3 + 5x^2 + 18x + 9\right) \div \left(x + 3\right)$

5. $\left(x^3 - 8\right) \div \left(x - 2\right)$

Additional Topics
Topic A.6 Complex Numbers

Vocabulary
imaginary numbers   •   complex numbers   •   powers of i

1. The set of _____ consists of numbers of the form bi where b is a real number, and $b \neq 0$.

2. _____ are numbers of the form $a + bi$, where a and b are real numbers.

Step-by-Step Video Notes
Watch the Step-by-Step Video lesson and complete the examples below.

Example	Notes
1 & 2. Simplify. $$\sqrt{-49}$$ $$\sqrt{-8}$$	
3 & 4. Evaluate the powers of i. $$i^9$$ $$i^{15}$$	
5. Add or subtract. $$(5 + 6i) + (6 - 3i) = \Box + \Box + \Box - \Box$$ $$= (\Box + \Box) + (\Box - \Box)$$ $$= \Box + \Box$$ Answer:	

Example	Notes
8. Multiply. $(7-6i)(2+3i)$ Answer:	
10. Divide. $\dfrac{7+i}{3-2i}$ Answer:	

Helpful Hints

For all positive real numbers a, $\sqrt{-a} = \sqrt{-1} \cdot \sqrt{a} = i\sqrt{a}$.

Note that $i^{4n} = 1$ where $n \neq 0$. Keep this in mind when evaluating powers of i.

Concept Check

1. Why is $1 + \pi \cdot i^4$ a real number while $\pi + i^5$ is not? Are they complex numbers? If so, what are their real and imaginary parts?

Practice

Add, subtract, multiply, or divide as indicated.

2. $(1.28 + 2i) + (11 - 5i)$

3. $\left(\dfrac{3}{2} + 8i\right) - (3 - 6i)$

4. $(6 - 2i)(2 + 7i)$

5. $\dfrac{1 + 4i}{7 - 3i}$

Additional Topics
Topic A.7 Matrices and Determinants

Vocabulary
matrix • dimensions of a matrix • determinant • value of a 2×2 determinant
value of a 3×3 determinant • minor of an element of a 3×3 determinant • array of signs

1. A _____ is a rectangular array of numbers that is arranged in rows and columns.

2. The _____ $\begin{vmatrix} a & c \\ b & d \end{vmatrix}$ is $ad - bc$.

Step-by-Step Video Notes
Watch the Step-by-Step Video lesson and complete the examples below.

Example	Notes
1. Write the coefficients of the variables in the system of equations as a matrix. $$\begin{aligned} 5y &= 10 \\ 2x - 6y &= 17 \end{aligned} \qquad \begin{bmatrix} \square & \square \\ \square & \square \end{bmatrix}$$	
3. Find the determinant of the matrix. $$\begin{bmatrix} 0 & -3 \\ -2 & 6 \end{bmatrix}$$ Answer:	
4. Evaluate the determinant. $$\begin{vmatrix} 4 & 1 & 2 \\ 3 & -1 & 0 \\ 1 & 2 & 3 \end{vmatrix}$$ Answer:	

Example	Notes
7. Evaluate the determinant by expanding it by minors of elements in the first column. $\begin{vmatrix} 2 & 3 & 6 \\ 4 & -2 & 0 \\ 1 & -5 & -3 \end{vmatrix}$ Answer:	

Helpful Hints

To evaluate a 3×3 determinant, use expansion by minors of elements in the first column.

The minor of an element (number or variable) of a 3×3 determinant is the 2×2 determinant that remains after we delete the row and column in which the element appears.

Concept Check

1. What does the array of signs help to determine?

Practice

Evaluate the determinants.

2. $\begin{vmatrix} 4 & 0 \\ 10 & -1 \end{vmatrix}$

3. $\begin{vmatrix} 3 & 4 \\ -1 & -2 \end{vmatrix}$

Evaluate the determinant by expanding by minors.

4. $\begin{vmatrix} 1 & 3 & 4 \\ -2 & 0 & 0 \\ -4 & 5 & -1 \end{vmatrix}$

Additional Topics
Topic A.8 Cramer's Rule

Vocabulary

Cramer's rule   •   value of a 2×2 determinant   •   system of linear equations

1.   _____ states that the solution to $\begin{array}{l} a_1 x + b_1 y = c_1 \\ a_2 x + b_2 y = c_2 \end{array}$ is $x = \dfrac{D_x}{D}$ and $y = \dfrac{D_y}{D}$,

$D \neq 0$, where $D = \begin{vmatrix} a_1 & b_1 \\ a_2 & b_2 \end{vmatrix}$, $D_x = \begin{vmatrix} c_1 & b_1 \\ c_2 & b_2 \end{vmatrix}$, and $D_y = \begin{vmatrix} a_1 & c_1 \\ a_2 & c_2 \end{vmatrix}$.

Step-by-Step Video Notes

Watch the Step-by-Step Video lesson and complete the examples below.

Example	Notes
1. Solve the system by Cramer's rule. $-3x + y = 7$ $-4x - 3y = 5$ Find D. $D = \begin{vmatrix} \Box & \Box \\ \Box & \Box \end{vmatrix} = (\Box)(\Box) - (\Box)(\Box) = \Box$ Find D_x. Find D_y. Solve for x and y. $x = \dfrac{D_x}{D} = \dfrac{\Box}{\Box} = \Box$ and $y = \dfrac{D_y}{D} = \dfrac{\Box}{\Box} = \Box$ Answer:	

Example	Notes
3. Solve the system by Cramer's rule. $2x - y + z = 6$ $3x + 2y - z = 5$ $2x + 3y - 2z = 1$ Answer:	

Helpful Hints

$$a_1 x + b_1 y + c_1 z = k_1$$

The solution to $a_2 x + b_2 y + c_2 z = k_2$ is $x = \dfrac{D_x}{D}$, $y = \dfrac{D_y}{D}$, and $z = \dfrac{D_z}{D}$, $D \neq 0$, where

$$a_3 x + b_3 y + c_3 z = k_3$$

$$D = \begin{vmatrix} a_1 & b_1 & c_1 \\ a_2 & b_2 & c_2 \\ a_3 & b_3 & c_3 \end{vmatrix}, \ D_x = \begin{vmatrix} k_1 & b_1 & c_1 \\ k_2 & b_2 & c_2 \\ k_3 & b_3 & c_3 \end{vmatrix}, \ D_y = \begin{vmatrix} a_1 & k_1 & c_1 \\ a_2 & k_2 & c_2 \\ a_3 & k_3 & c_3 \end{vmatrix}, \text{ and } D_z = \begin{vmatrix} a_1 & b_1 & k_1 \\ a_2 & b_2 & k_2 \\ a_3 & b_3 & k_3 \end{vmatrix}.$$

The value of the 2×2 determinant $\begin{vmatrix} a & b \\ c & d \end{vmatrix}$ is $ad - bc$.

Concept Check

1. Would you solve the system $\begin{array}{l} x + y = 1 \\ -x + y = 3 \end{array}$ using Cramer's rule? Explain why or why not.

Practice

Solve the systems by Cramer's rule.

2. $\quad 3x = 9$
   $\quad -8x + 10y = -4$

3. $\quad 4x + 6y = 48$
   $\quad -x + 2y = 16$

4. $\quad x + 2y + z = 5$
   $\quad -x + 4y - 2z = 12$
   $\quad -3x + y = 9$

Additional Topics
Topic A.9 Solving Systems of Linear Equations Using Matrices

Vocabulary
augmented matrix   •   matrix row operation

1. A matrix that is derived from a system of linear equations is called the _____ of the system.

Step-by-Step Video Notes
Watch the Step-by-Step Video lesson and complete the examples below.

Example	Notes
1. Solve the system of equations using a matrix. $4x - 3y = -13$ $x + 2y = 5$ Write the augmented matrix and perform row operations to solve the system of equations. $R_1 \leftrightarrow R_2$ $-4R_1 + R_2$ $-\frac{1}{11}R_1$ Write the equivalent system and solve. Answer:	

Example	Notes
2. Solve the system of equations using a matrix. $2x + 3y - z = 11$ $x + 2y + z = 12$ $3x - y + 2z = 5$	

Answer:

Helpful Hints

An augmented matrix is made up of two smaller matrices separated by a vertical line. The coefficients of the variable terms are placed to the left of the vertical line, and the constant terms are placed to the right.

In a matrix, all the numbers in any row or multiple of a row may be added to the corresponding numbers of another row. All the numbers in a row may be multiplied or divided by any nonzero number. Any two rows of a matrix may be interchanged.

Concept Check

1. Can you think of a situation where solving a system of linear equations using matrices would not be the best method? Explain.

Practice

Solve the systems of equations using matrices.

2. $-2x + 3y = 2$
   $-4x + y = -16$

3. $4x - 2y = 7$
   $3x + 2y = 7$

4. $x + 4y - 5z = 4$
   $3x - 8y + 7z = 9$
   $2x + 9z = 17$

Name: _____   Date: _____

Instructor: _____   Section: _____

Additional Topics
Topic A.10 Basic Probability and Statistics

Vocabulary
probability • favorable outcomes • possible outcomes

1. _____ measure(s) the likelihood that a given event will occur.

Step-by-Step Video Notes
Watch the Step-by-Step Video lesson and complete the examples below.

Example	Notes
2 & 3. Find the following probabilities based on rolling a six-sided die. Find the probability of rolling a 4. $$P\left(\square\right) = \frac{\square}{\square}$$ Find the probability of rolling a 1 or a 3. $$P\left(\boxed{}\right) = \frac{\square}{\square} = \frac{\square}{\square}$$	
4. If a coin is tossed twice, find the probability of throwing tails twice in a row. Begin by drawing a tree diagram. Use the tree diagram to determine the probability. Answer:	

Example	Notes
6. If a die is tossed and then a coin is tossed, find the probability of rolling a 4 and then tossing a head.	
Answer:	
8. A bag contains 6 red marbles, 7 blue marbles, 2 black marbles, and 5 green marbles. Find the probability of picking a blue or black marble.	
Answer:	

Helpful Hints

Probability can also be defined as $\text{Probability} = \dfrac{\text{Favorable Outcomes}}{\text{Possible Outcomes}}$.

When multiple events occur, the probability of each event is multiplied to get the probability for all the events.

Concept Check
1. Explain why probabilities can only be between 0 and 1.

Practice

A die is tossed and then a coin is tossed. Find the following probabilities.

2. The die lands on an even number and the coin lands on heads.

3. The die lands on a multiple of 3 and the coin lands on heads or tails.

A bag contains 4 red marbles, 7 green marbles, and 3 white marbles.

4. Find the probability of picking a black marble.

5. Find the probability of picking a red, green, or white marble.

Example	Notes
6. If a die is tossed and then a coin is tossed, find the probability of rolling a 4 and then tossing a head. Answer:	
8. A bag contains 6 red marbles, 7 blue marbles, 2 black marbles, and 5 green marbles. Find the probability of picking a blue or black marble. Answer:	

Helpful Hints

Probability can also be defined as $\text{Probability} = \dfrac{\text{Favorable Outcomes}}{\text{Possible Outcomes}}$.

When multiple events occur, the probability of each event is multiplied to get the probability for all the events.

Concept Check
1. Explain why probabilities can only be between 0 and 1.

Practice

A die is tossed and then a coin is tossed. Find the following probabilities.

2. The die lands on an even number and the coin lands on heads.

3. The die lands on a multiple of 3 and the coin lands on heads or tails.

A bag contains 4 red marbles, 7 green marbles, and 3 white marbles.

4. Find the probability of picking a black marble.

5. Find the probability of picking a red, green, or white marble.